Playing
with
Reality

Playing
with
Reality

How Games Have

Shaped Our World

Kelly Clancy

Riverhead Books

New York

2024

RIVERHEAD BOOKS
An imprint of Penguin Random House LLC
penguinrandomhouse.com

LIBRARY OF CONGRESS CONTROL NUMBER: 2024935063

ISBN 9780593538180 (hardcover)
ISBN 9780593538203 (ebook)

Printed in the United States of America
1st Printing

Book design by Cassandra Garruzzo Mueller

To the next generation—especially to my son,
Cillian, and my niblings Fiona, Kira, and Ronan.
If play is the engine of creation,
I cannot wait to see what new worlds you build.

Contents

Part I

How to Know the Unknown

1

The Play of Creation

Lila is the play of creation. To awakened consciousness, the entire universe, with all its joys and sorrows, pleasures and pains, appears as a divine game, sport, or drama. It is a play in which the one Consciousness performs all the roles.

KENDRA CROSSEN BURROUGHS

The long-forgotten game rithmomachia was once medieval Europe's most popular educational technology. It remained a standard in the monastic curriculum for nearly five hundred years. Ovid hailed it as "the leaf, flower, and fruit of Arithmetic, and its glory, laud, and honor." Thomas More depicted the virtuous citizens of his *Utopia* playing rithmomachia instead of "ruinous" games like dice. Elites played it, in part, to demonstrate their erudition. Church leaders believed that the game was not only an educational force but also an enlightening one: it instructed players in the greater harmony of the universe and even tamed violent tempers. Rithmomachia, a "banquet proper of the mind," wordlessly introduced students to divine truths through the beauty of play.

Also known as the Philosopher's Game, rithmomachia was a capture game related to chess. Two players moved black and white pieces across a checkered board according to rules prescribed by their respective shapes. Unlike in chess, pieces were engraved with numbers meant to represent ideas in number theory. A player could capture an opponent's piece only if the numbers inscribed on the battling pieces had some geometric relation. A player won by arranging three or four pieces on their opponent's side of the board into a "harmony," or geometric progression.

The game has been traditionally credited to the fifth-century BCE Greek philosopher Pythagoras, though the earliest written evidence of its existence dates only to 1030 CE. Nevertheless, it was clearly designed to reflect his teachings. Rithmomachia was considered the apotheosis of the quadrivium, a philosophy of higher education that embodied Pythagorean ideals, centered around his worship of numbers. The quadrivium consisted of arithmetic (pure numbers), geometry (numbers in space), music (numbers in time), and astronomy (numbers in space and time). In rithmomachia, numbers and shapes danced across the board, organized into spatial patterns unfolding over time. These ideas, harmless as they seem, laid the foundations for one of the most absurd cover-ups in intellectual history.

Though none of Pythagoras's writings survived the march of history, he left an outsize influence on European thought. Born in Greece, he was rumored to have spent twenty years studying with religious mystics in Egypt before establishing a cult in the Grecian colony of Croton. His was an ascetic religion: he preached vegetarianism and temperance in an era of decadence and growing political instability. His followers shared all their property and ate their meals together. Pythagoras forbade his disciples from eating beans, which he fretted might house the souls of men, given their resemblance to human fetuses. Above all, his teachings held numbers sacred: they were the building blocks for the entire cosmos. Numbers were beyond divine— they were the very cause of gods.

"All is number," Pythagoras professed. Yet not all numbers were created equal, in his view. Central to his cosmology was the idea of harmony. He had discovered that musical pitch varies depending on a plucked string's length, and musical harmony arises from plucking strings that are precise multiples of one another. For this reason, he held ratios sacred, because he believed they expressed the harmony of the universe. He worshipped what are now known as rational numbers, including whole numbers (like one, two, and three) and their ratios (like one-third and five-sixths). These fundamental elements, he believed, comprise all other phenomena in the universe.

His growing cult gained political clout over the two decades that it resided in Croton. Eventually, local leaders deemed the bizarre vegetarian a threat to their power. They organized an angry mob to drive the cult away, burning the commune to the ground. Different legends have Pythagoras meeting his end by exile, fire, or suicide. One story has him fleeing for his life, only to be stopped in his tracks by a field of beans. Unwilling to trample across the sacred crop, Pythagoras was caught and executed by his pursuers. Despite this fall from grace, his cult persisted for another three hundred years, and his ideas dominated European philosophy and education for nearly two millennia.

Dogma is anathema to intellectual progress. Unfortunately, Pythagoras's devotion to rational numbers motivated a scholarly subterfuge. Hippasus, a member of Pythagoras's cult, is thought to have accidentally discovered irrational numbers in the fourth century BCE. Whereas rational numbers can be expressed as the ratio of two numbers, irrational numbers can't be. Rational numbers can be precisely determined: One-half is equal to 0.5. The value of one-third is 0.3333 . . . , where the three repeats forever. We automatically know that its seven-billionth digit is three. The square root of two, on the other hand, is irrational. It equals 1.41421356237 . . . , a decimal without repeating or terminating digits. Irrational numbers can't be precisely determined. To know the seven-billionth digit of the square root of two, we must calculate it out. Irrational numbers, Hippasus realized, are alogos, or "inexpressible." Their existence directly contradicted the Pythagorean doctrine that all phenomena are composed of whole numbers and their ratios. They contravened all the certainty and comprehensibility that Pythagoras believed numbers endowed upon the universe. Hippasus disappeared into history shortly after sharing his discovery; he was reportedly executed at sea. Legend has it that the cult justified his murder as an offering to Poseidon, rendering it an act outside legal jurisdiction.

Some medieval historians blame the quadrivial curriculum for holding European scholarship back hundreds of years. Its focus on harmony blinkered scientists to fundamental advances, such as the discovery of irrational numbers and celestial mechanics. The quadrivial curriculum was largely abandoned by the seventeenth century, as new mathematical techniques from Persia and India led to breakthroughs in calculus and probability theory. Rithmomachia similarly fell by the wayside. The harmonies that Pythagoras and his followers believed made up the material cosmos were flimsy ideals. Planets move along elliptical, not spherical, orbits, defying models of cosmic conformity. Not only do irrational numbers exist, but the size of their set is so huge it utterly swamps that of rational numbers. If one were to write down all rational and irrational numbers on slips of paper and place these into an immense hat, one would have almost no chance of picking out a rational number—they're vanishingly rare in the universe of mathematics.

Games are a kind of untrue truth. What is internally consistent within a game need not reflect anything about reality. Yet games have been increasingly adopted as models of the world. Games are a method by which we can learn how to reason about the behavior of objects in rule-based systems. At their best, games have generated a deeper understanding of mathematics. At the same time, ideologues have used games to smooth reality's rougher edges.

To be fair, the problem is never a game itself but the dogma it's co-opted to serve. It was the doctrine of harmony, not rithmomachia, that held thinkers back. Rithmomachia was the doctrine's emissary, imprinting generations of scholars with its aesthetic ideals. A game rewards its players for adopting its precepts. A pacifist cannot win *Call of Duty* without shooting in-game enemies; a socialist cannot win Monopoly without adopting capitalist behaviors. Similarly, rithmomachia seduced its players with the beauty of Pythagorean ideals. Games are more than models of the world—they're models that reward us for believing in them. They can influence how we think of and see the world, for better or worse. As such, games are also a window into the beliefs and habits of the people who play them. By examining our relationship with games throughout history, we can better understand the beliefs of people long past and more clearly see our own.

Games are older than written language. Games like Go, chess, backgammon, and mancala are living artifacts; they've outlasted empires and conquered cultures. They transcend even language: through play, we can interact with other people's minds, regardless of their tongues. One can imagine an ancient Mesopotamian tavern populated by far-flung merchants and travelers, unable to converse but still able to enjoy an evening together engrossed in a board game. Throughout history, immigrants carried their treasured games with them and kept them alive for thousands of years, as cherished as any cultural memento. Today, multiple subcultures spanning continents and generations define themselves through the games they play. Games endure because they engage a learning system common to all people. They're a product of both culture and biology—a stimulus the brain devised, over time, to serve itself free pleasure. Though often dismissed as "trivial," games have captivated human attention for millennia.

People are so entranced by games that they've routinely used them to anesthetize the anxiety of troubled times. In his account of the Greco-Persian Wars, Herodotus contends that gaming saved the Lydian people in a time of deprivation. The Lydians were much like the Hellenes, he writes. They were the first people to mint coins and, by their own account, the first to invent several now ubiquitous games. Around the second millennium BCE, they were devastated by a decades-long famine but found solace in play:

> When they saw the evil still continuing, they sought for remedies, and some devised one thing, some another; and at that time the games of dice, knucklebones, ball, and all other kinds of games except draughts, were invented, for the Lydians do not claim the invention of this last; and having made these inventions to alleviate the famine, they employed them as follows: they used to play one whole day that they might not be in want of food; and on the next, they ate and abstained from play; thus they passed eighteen years.

Though the Lydians couldn't have invented the games they claimed to have invented—knucklebones and dice far predate them—it's conceivable that they took refuge in games when food was scarce. Games are, by design, utterly absorbing. History is littered with stories of famous figures ruined by their love of games or gambling. To this day, a handful of people die every year while gaming—usually from exhaustion, though some are murdered in the heat of competition. This is not cited to demonize games but to demonstrate their power. For thousands of years, games have fascinated people. Today, games are arguably the dominant cultural media. They are a superstimulus, a psychotropic technology, having been tweaked over time to maximally engage our reward systems. By examining something that humans love, we can better understand humans. What's more, we can better understand understanding: games have shaped how we generate knowledge and reason about the unknown.

One difficulty in discussing games is their elasticity as a metaphor. Is make-believe a game? Is an exam a game? Is a puzzle? There are zero-player games, massively multiplayer online games, and games with negative-, zero-, or positive-sum rewards. To avoid lengthy classifications, I'll roughly define a game as a system furnished with a goal. Players make decisions given some uncertainty (whether the roll of a die or the strategy of an opponent) in pursuit of that goal. Play is all about the unknown and learning how to navigate it. "Fun is just another word for learning," writes game designer Raph Koster. Dice and chess have little in common except that players work to predict the world in both cases. In gambling, players hope to anticipate the environment. In chess, they hope to anticipate their opponents. A game might look like an auction (with the goal of acquiring a coveted good at a reasonable price), social media (with the goal of attracting attention), or *SimCity* (with the goal of designing a sustainable digital metropolis).

Chance is central to many games because randomness is nature's foundational search algorithm. Random mutations are the driving engine of evolution. Modern scientific experiments incorporate randomization to eliminate the influence of scientists' personal choices and biases. Machine-learning engineers use randomness in their models to jolt the system out of local minima and explore the solution space more thoroughly. It's also at the heart of our ancient drive to play. Before humans, before mammals, this drive emerged with the earliest animals.

Play expanded nature's search strategy of randomness into the behavioral realm. It's been crucial to the emergence of intelligence. Evolution responds to changes in the environment on relatively slow timescales. Innovation happens only as fast as genetic mutations can spread through a population. The development of the nervous system allowed animals to rapidly respond to their environments—to migrate when the climate changes, say, or learn how to avoid an invasive poisonous species. Hardwired reflexes can be maladaptive, however: take, for instance, many cats' absurd overreaction when they suddenly see a cucumber, triggered by their instinctual fear of snakes. Play uncouples behavior from rote instinct, facilitating flexibility. It draws chance into the realm of experience; it's a safe platform to test the unknown. Play opens new realms for animals, allowing them to build an inventory of adaptable, robust behavioral programs. In play, animals develop a range of strategies through random exploration: If condition A, try B. If condition C, try D, then E. In place of the reactive response of a hardwired reflex, play enables animals to explore a suite of options, some more adaptive than others. It is the crucible of invention and a learning system that mimics the genius of evolution. In natural selection, mutations resulting in bodies that meet the demands of their environments are rewarded with survival. In play, behaviors that correctly anticipate the demands of the environment persist. Play is to intelligence as mutation is to evolution.

This is also why play so often involves being out of control in some way. Otters sled down muddy hills; birds surf on the wind; wrestling children tumble. Play is practice for the unexpected. Playing animals get themselves into purposeful but safe trouble, a strategy that helps them learn how to best escape real trouble. Play allows our brains to build more robust models of the world, sampling from experiences we wouldn't otherwise encounter. It's also how we develop and stress-test social relationships. Play is an activity that brings the wilderness of reality into the realm of the understood.

While play is how animals explore their bodily agency, games are how people explore their mental agency. Games have helped humans improve their powers of reasoning and interactive decision-making for millennia. The first evidence of game boards dates to about ten thousand years ago, just as cats were beginning to be domesticated and when agriculture was still an emerging technology. Limestone boards carved with rows of depressions were common in Neolithic households of the Near East, potentially used for a counting game similar to mancala. The game may have helped players understand basic mathematics. By manipulating pebbles for fun, they would have grown more familiar with the abstraction we now call numbers.

Beyond the pleasure they bring us, games have reshaped how we think. Games are mental practice, empowering players to develop cognitive skills like strategy, numeracy, and theory of mind in a safe domain. Chess, Go, and related games develop social and strategic thinking. Language and knowledge games like trivia, Scrabble, and puns develop mnemonic abilities and literacy. Games of chance help people develop counting skills and probabilistic estimation. Ninth-century Chinese printmakers invented the first playing cards, founding games of incomplete information. Players, unable to see one another's cards, have to reason about their opponents' psychology and intentions. These games sharpen our understanding of each other. Games, says AI researcher Julian Togelius, are like casts of the mind. Each is fit to a different cognitive capability, and new games emerge as we identify new mental niches and functions. Games like *Pokémon* scratch our itch to collect things, while games like *Tetris* satisfy our compulsion to organize things. Game design is a cognitive science. Games themselves are like a "homebrew neuroscience," says game designer Frank Lantz, "a little digital drug you can use to run experiments on your own brain." We see ourselves—our biases, weaknesses, strengths—more clearly through play.

In life, we often learn the rules of the world by observing the consequences of our actions. For example, if we touch a hot stove, we learn that it's painful and that we should avoid touching it in the future. We infer rules ("don't touch hot stoves") from consequences ("ouch, that hurt"). In games, consequences are determined by rules. Players must reason forward about the ramifications of the decisions they make. In order to infer the effects of their choices, chess players must understand the rules behind how pieces interact. Games train us to explicitly reason about effects—a crucial mental skill for humans living in complex societies mediated by rules.

In fact, as we'll continue to explore in this book, games are at the very heart of how we conceive of understanding itself. The late physicist Richard Feynman expounded on this in his famous Caltech course, which he taught in the early 1960s. "What do we mean by 'understanding' something?" he asked of the crowded lecture hall. We can say we understand a system if we know the rules governing it:

We can imagine that this complicated array of moving things which constitutes "the world" is something like a great chess game being played by the gods, and we are observers of the game. We do not know what the rules of the game are; all we are allowed to do is to watch the playing. Of course, if we watch long enough, we may eventually catch on to a few of the rules. The rules of the game are what we mean by fundamental physics. Even if we knew every rule, however, we might not be able to understand why a particular move is made in the game, merely because it is too complicated and our minds are limited. If you play chess, you must know that it is easy to learn all the rules, and yet it is often very hard to select the best move or to understand why a player moves as he does. [. . .] We must, therefore, limit ourselves to the more basic question of the rules of the game. If we know the rules, we consider that we "understand" the world.

The rules of a system (or game) are its most efficient representation. We can't hope to know every possible move in a game. The enormous space of all legally possible Go games (somewhere between 10^{800} and a googolplex, or $10^{10^{100}}$, unfolds as a direct consequence of the game's three fundamental rules, playing out over a nineteen-by-nineteen grid. Once we know the rules well enough to predict or explain a move, we can claim to understand the game, even without elaborating all its possible outcomes. Prediction, as we'll see, is the main currency of the brain. It feels rewarding to have a predictive model of the world. Yet knowing the rules of a system doesn't always mean we can anticipate their outcomes. Incredibly complex dynamics can arise out of deceptively simple rules. Game models can manufacture the illusion that we understand a system better than we do, echoing Pythagoras's obsession with rational numbers obscuring the complexities of material reality.

Games have endured because they represent a model of how our minds work. They're the outgrowth of a learning system that has been central to the evolution of intelligence. Play is a tool the brain uses to generate data on which to train itself, a way of building better models of the world to make better predictions. Perhaps this is also why games have been traditionally associated with divination: throughout history, people have harbored an enduring intuition that games are somehow linked to knowing the future. Cards, dice, and lots have long been used as decision-making tools—an instinct that, as we'll discover, helped to de-bias human choices. The notion of rules and their consequences sits at the root of what we mean by "understanding." None of this is an accident. Games are more than an invention; they are an instinct.

Games have been a powerful generator of knowledge. Though rithmomachia was saddled with Pythagorean dogma, other games yielded profound mathematical insights. Their contemplation gave rise to probability theory and modern economic thought, as well as new ideas in moral philosophy and artificial intelligence. In contrast to

Pythagoras and his cover-up of irrational numbers, the mathematician John Conway discovered an immense new universe of numbers, now known as the surreal numbers, in 1974, after contemplating the endgames of Go. This was the largest magnitude of infinity that mathematicians had discovered in a century. Conway was stunned by the immensity of his revelation. It put him in a stupor for weeks—he felt as though he'd stumbled upon a new continent.

Even so, games are ultimately mathematical objects. We can use them to generate knowledge about simulated worlds, but this knowledge doesn't necessarily translate to reality. The orderly randomness of dice is a poor model for the untamed randomness of the real world. This discrepancy contributed to the 2008 global financial crash because traders failed to incorporate appropriate measures of risk into their bets. Game theory, once an arcane corner of pure mathematics, now serves as the foundation of modern economics—despite being a poor model of real people. This hasn't stopped companies and academics from employing game theory in the design of economic and political systems that underpin our everyday lives.

We should be especially careful about using games as a lens for understanding people, because the metaphors we use to understand ourselves matter. In defending his controversial theories about human behaviors, the twentieth-century psychologist B. F. Skinner claims that "no theory changes what it is a theory about; man remains what he has always been." Refuting this claim lies at the heart of this book. It's certainly true in realms like physics: atoms survived millennia unscathed by our early, incorrect theories about them. An electron's orbit is entirely determined by the attractive and repulsive physical forces buffeting it, untouched by our models. But people aren't passive objects of physics. People, unlike atoms, learn. They are independent agents who make choices dependent on their beliefs about the world and how it works—and their beliefs about themselves. As the engineer

Edsger W. Dijkstra writes, "The tools we use have a profound (and devious!) influence on our thinking habits, and, therefore, on our thinking abilities."

At the individual level, a person who believes humans to be nasty and brutish will make different choices in life than one who believes that human nature is gentle and generous. Their beliefs affect their politics, their priorities, their prejudices. At the societal level, an incorrect model of people, put into practice in the economic spheres they inhabit, can warp their behaviors and radically change their experience of life. A game rewards its players for adopting the precepts from which it's built. Its rules dictate whether players should cooperate or compete, play honestly or cheat. Game theorists and game designers increasingly shape our shared civic circles, and we're rewarded for adopting their models of humanity in games we cannot escape. It's never been more critical to examine how games have come to dominate modern thought, and how we might clarify our thinking from their rule.

2

How Heaven Works

To learn is a natural pleasure, not confined to philosophers, but
common to all men.

ARISTOTLE

n the early twentieth century, a fever swept through the world, has-
tened by armies fanning out across Europe during World War I. Previously unknown to science, it was popularly deemed "sleeping
sickness." People across the world fell into lethargy, sometimes instan-
taneously. A British doctor recalled the first case he encountered: a
healthy girl walking home from a concert simply folded in half and fell
to the ground. Within half an hour, her sleep was so deep that she
could no longer be roused; within twelve days, she was dead.

Encephalitis lethargica, as it's now known, is a disease without a fixed
character. Its elastic nature is reflected in the diversity of its symptoms.
With its most common form, patients are overwhelmed by the desire to
sleep, though even in this pseudosleep they remain dimly aware of what
is happening around them. Some patients recover. Others progress to a
chronic form of the disease, characterized by a motley collection of
symptoms: euphoria, increased sexual drive, "excessive" pun making,
tremors, muscle rigidity, hallucinations, and self-mutilation. An eight-

year-old girl pulled out all her teeth and plucked out both of her eyes. A seventeen-year-old boy became obsessed with horrible odors. He sought out armpits and feces and hoarded trash in his room. Some patients progress to paralysis, coma, or death. Patients languish in long-term care for decades, still as statues, stone-faced and trapped in perpetual slumber. The cause of the disease remains unknown to this day. Its effects, doctors later found, could be mitigated with a then obscure chemical now known as dopamine.

The neurotransmitter dopamine suffers one of the most persistent branding problems of any chemical. This is much to its own credit: it simply does so much. Today, it's commonly mischaracterized as hedonism's biological counterpart, the "pleasure molecule." But for decades after its discovery, scientists thought little of it. It was first synthesized in 1910 as an intermediary in the production of adrenaline, which was then used as an asthma medication. Dopamine continued to be found, here and there, in various bodily tissues. Still, it was dismissed as nothing more than a way station for the chemicals, like adrenaline, that were thought to matter. A traditional Ayurvedic medicine would soon upend this.

In India, high blood pressure, fevers, and insanity had been treated for centuries with the flowering shrub sarpagandha. Mahatma Gandhi, who suffered from hypertension, devoutly took six drops of sarpagandha tincture in his tea daily. Sarpagandha was later "discovered" in the early 1950s by the American doctor Robert Wilkins, after having been studied in clinical trials by Indian scientists for over a decade. Chemists isolated its active compound, reserpine, which became a popular blood-pressure medication and antipsychotic. It was also sometimes used as an animal tranquilizer: in high doses, reserpine rendered animals catatonic. No one knew precisely how.

In 1957, the Swedish researcher Arvid Carlsson and some of his colleagues injected mice with reserpine. They found that it depleted the animals of several chemicals, including dopamine and its down-stream products, such as noradrenaline. Noradrenaline was then known for its effect of arousing the body to action. The scientists therefore expected that injecting noradrenaline back into the catatonic animals would restore their ability to move. It did not, but injecting them with L-dopa, a precursor to dopamine, did. In some cases, the animals even grew hyperactive. This, along with Katharine Montagu's 1957 discovery of dopamine in brain tissue, definitively established dopamine's role as a neurotransmitter—a finding that earned Carlsson a Nobel Prize. The notion that neurons could talk to one another through bulk chemical transmission was still baffling. At the time, electrical signaling, in which individual cells send messages to select partners, was the more familiar mode of neural communication. Here, a single message encoded in the concentration of dopamine was being broadcast to entire brain areas. But what was it saying?

Oleh Hornykiewicz, a Viennese neurologist, noticed that the effects of dopamine depletion in animals echoed the symptoms of the common neurodegenerative disorder Parkinson's disease. Parkinson's was first described in 1817 by the surgeon James Parkinson, whose patients displayed "involuntary tremulous motion, with lessened muscular power, in parts not in action and even when supported; with a propensity to bend the trunk forward, and to pass from a walking to a running pace: the senses and intellects being uninjured." Parkinson, an avid fossilist and naturalist, described it as a new "species of disease," one that he'd drawn out from the jungle of nebulous neurological symptoms, like a botanist classifying a newly discovered flower.

Though Parkinson's is most common in elderly patients, it was also frequently seen in people with chronic sleeping sickness. In the years following the original sleeping sickness epidemic, seemingly catatonic patients flooded care facilities. In the 1960s, more than thirty years after the first epidemic, Hornykiewicz began collecting the brains of recently deceased patients. He discovered that the brains of patients with parkinsonism were low in dopamine. What if he could reverse parkinsonism with L-dopa, just as Carlsson had revived catatonic mice? He wasted no time and immediately delivered his store of the drug to a colleague in charge of an elder-care facility in Vienna. The medical staff administered L-dopa to Parkinson's patients. The results seemed miraculous. Patients who had been immobile for decades, imprisoned by sleeping sickness, stood up and walked, their voices and personalities returned to them.

Dopamine replacement therapy became the standard of care for Parkinson's patients, as it remains to this day. It's not a cure, unfortunately; it becomes less effective over time. In his breakout work, *Awakenings*, the neurologist Oliver Sacks recounted the brief but luminous revival of encephalitis lethargica patients, likening them to "extinct volcanoes" erupting back into life. With this spectacular success, dopamine rocketed into the scientific spotlight, becoming one of the most researched neurotransmitters. Scientists have since discovered dopamine in almost every animal with a nervous system, making it evolutionarily ancient; it's involved in the movement of flatworms, fireflies, flounder, and falcons alike. All of this seemed to cement dopamine's place as the neurotransmitter responsible for movement, but nature is never so simple.

- - - - -

No sooner had the computer been conceived than people wondered whether it could think. The world's earliest programmer, Ada Lovelace, was the first to realize that Charles Babbage's protocomputer, the Analytical Engine, could do more than crunch numbers. It might one day make music, prove math theorems, play games. Yet, she argues, it could only carry out instructions; it had "no pretensions whatever to *originate* anything. It can do whatever we *know how to order it* to perform." A century later, Alan Turing, whose work led to the creation of more powerful and flexible computers, believed that these machines would one day be capable of much more. He responded to "Lady Lovelace's Objection": "A better variant of the objection says that a machine can never 'take us by surprise.' [. . .] Machines take me by surprise with great frequency." He predicted that computers would eventually be able to generate new knowledge and understanding. But first we'd have to teach them how to learn.

Designing a machine that captured all the complexity and knowledge of a fully formed adult, Turing anticipated, would be difficult. Its intelligence might simply be credited to its creators—it would not, as Lovelace proclaimed, originate anything. "Instead of trying to produce a programme to simulate the adult mind," Turing writes, "why not rather try to produce one which simulates the child's? If this were then subjected to an appropriate course of education, one would obtain the adult brain." He imagined that this "child" machine could be, like children, educated through punishment and reward. This entailed solving two separate problems. First, researchers would have to build a computer program that reflected a child's capacity to learn. Then they'd have to design its education. Today, we'd call the solution to the first problem a learning algorithm and the solution to the second its training data.

Turing suggested board games as an ideal training arena for bringing up these child machines. Games are miniature worlds, abstractions of interaction, and their discrete nature makes them beautifully suited to computers. In games like chess and checkers, a few simple rules unfold into campaigns of astronomical complexity. Games have long been thought to demonstrate their players' intelligence. A player's progress from novice to expert is conveniently assessed and tracked by metrics like their ranking. Thus, games could serve as both a training regimen for learning agents and a benchmark for measuring their intelligence. Structured games standardize players' agency, evening the playing field. Games impose symmetry. Players work against each other toward a shared goal: winning. They're given identical pieces and bound by the same rules. To measure time, we use clocks; to measure space, rulers. Games came to be adopted as a meter of intelligence. They are an ancient form of rhetoric—a debate not of words but of choices, actions parried across time and space.

John McCarthy, the researcher who coined the term *artificial intelligence*, also provided one of its most enduring definitions.* "Intelligence," he writes, "is the computational part of the ability to achieve goals in the world." We might assess a machine's intelligence through conversation, as in Turing's "imitation game," wherein players—both computer programs and real people—vie to convince a human judge that they're human via typed conversation. Or we might call a program that can beat a human in chess intelligent.

Turing was, by his own admission, a mediocre chess player. He nurtured his dream of building an artificially intelligent chess program in conversations with his favorite chess partner and colleague, Donald Michie, who was equally mediocre at the game. Both Turing and Michie worked at Bletchley Park during World War II, devising methods to decode encrypted Axis communications. Michie had come to Bletchley Park almost by accident, having enrolled in a cryptography class with hopes of doing "something unspecified but romantic" for the war effort. Today, we know that his research was instrumental to the Allied win: Michie's insights helped crack the Lorenz cipher and vastly improved the Colossus II computer. Thanks to his work, communications that

---

* I use the term *artificial intelligence* (AI) not because these programs are necessarily meaningfully intelligent themselves but because this is the aspiration of their creators.

had previously taken days to decode took mere hours, enabling Allied forces to dodge ambushes and anticipate enemy troop movements.

Michie was smitten with Turing's vision: "I resolved to make artificial intelligence my life as soon as it became feasible," he writes. But computers as a technology were vastly outpaced by researchers' ambitions for them. They were also impossibly expensive, and rare outside military circles. After the war ended, Michie returned to academia and shifted his focus to genetics, inspired by his boyhood love of mice. He was a middling biologist; his greatest contribution was to the research of his wife, Anne McLaren, whose work paved the way for in vitro fertilization. But Michie never gave up on the dream of artificial intelligence, despite having no access to computers.

In 1961, Michie took a bet with a colleague who was skeptical that machines could learn. Michie would win the bet using nothing more than three hundred-odd matchboxes and some tinted glass beads. He built a learning system that could play tic-tac-toe and named it MENACE: Matchbox Educable Noughts and Crosses Engine. Each matchbox represented a state of the tic-tac-toe board, and the boxes were stacked in piles representing sequential moves for all possible arrangements of X's and O's. The beads, nine colors in all, indicated every possible next move from the current state. Initially, Michie stashed an equal number of colored beads in each matchbox drawer. On every turn, he would randomly pull a bead from the matchbox, which determined MENACE's next move and the game board's next state. The color of the glass bead taken from the next box decided the next move, and so on. The drawers were left open as a record of which moves had been played. If MENACE lost at the end of the game, Michie didn't return the beads to the open drawers, lessening the probability that MENACE would take those actions in the future. If the game ended in a draw, he added one extra bead of the appropriate color to each matchbox. If the game resulted in a win, he returned three beads to each box. MENACE learned through reinforcement, with Michie rewarding correct moves and punishing incorrect moves.

At first, MENACE was terrible: "Random games have an extremely idiotic character, as can readily be verified by playing through one or two examples," Michie writes. But over the course of hundreds of games, the colored beads were redistributed within the boxes in a way that made winning moves more likely and losing moves less likely, much like deepening the ruts in the path leading to victory. Eventually, MENACE played tic-tac-toe perfectly. A mindless system had achieved expert performance through trial and error alone.

Michie had drawn inspiration from the trial-and-error theory of learning that had been the focus of psychology research earlier that century. The psychologist Edward Thorndike was determined to understand "animal stupidity"—how seemingly intelligent, purposeful behavior can be built from simple associations. He placed cats inside a puzzle box and put scraps of fish outside the box, just out of reach. The box had a trapdoor that opened only if its occupants pressed a lever. Once a cat's accidental step on the lever resulted in the door opening and a fish reward, the animals quickly learned, in subsequent trials, that they could escape the box by pressing the lever again. Thorndike named this the "law of effect": a behavior that leads to a pleasurable outcome will be selected and repeated. A behavior that leads to an unpleasant outcome will be extinguished. It's a bit like evolution, where gene variants that confer fitness are rewarded with survival in the population. In trial-and-error learning, a random action that leads to reward "survives." Instead of being recorded in DNA, it's stored in memory. The strength of its association with reward dictates how likely that action will be reproduced in the future. In MENACE's case, winning a game of tic-tac-toe ensured the survival of moves that led to the win. Victory "reproduced" winning moves, whereas defeat removed copies of losing moves from the pool of available actions.

Michie later applied similar learning methods to the endgames of chess. He was fond of a saying attributed to the Soviet mathematician Aleksandr Kronrod: chess was to the study of artificial intelligence as *Drosophila*, the fruit fly, had been to genetics. As studies of the simple *Drosophila* genome, consisting of only four chromosomes, had paved the way for understanding more complex human genetics, Michie pronounced that "the use of chess now as a preliminary to the knowledge engineering and cognitive engineering of the future is exactly similar, in my opinion, to the work on *Drosophila*."

Meanwhile, in the US, engineer Arthur Samuel had been working on a program that could play checkers. He'd originally conceived the project as a quick stunt to drum up funding to finish work on a computer he was building, not realizing that it would become a focus of his research for the next thirty years. Samuel wasn't a particularly strong checkers player, and his early efforts to build a checkers engine only performed as well as he could program it. To create a system that could exceed his own limited talents, he took up Turing's goal of building a program that could learn for itself, popularizing the phrase *machine learning* in 1959.

By the late 1950s, Samuel had struck on a game training regimen that would become central to the field of AI: self-play. In self-play, a program is trained against copies of itself. These copies tweak their parameters after each game to improve their win rates. Self-play has been so successful in part because players learn best from well-matched opponents. When a program is pitted against a copy of itself, it's always met by an adversary playing at its own level. If paired with a much better player, it might never learn, because it will always get creamed. If paired with a weaker opponent, it might win so easily that it's never challenged to improve.

However, training by self-play requires copious training time—a luxury few researchers had, given that computers were so rare. By then, Samuel was working at IBM, whose executives were not particularly enthused by his checkers side hustle. After his colleagues had gone home and the company's computers stood idle, Samuel snuck into the lab to train his program every night from midnight to seven in the morning. By 1956, the program was good enough to contend with novice players. Impressed by Samuel's furtive progress, the president of IBM arranged a public demonstration of the program, and the price of IBM stock jumped fifteen points overnight.

- - - - -

At the same time that computer scientists were attempting to engineer intelligent systems, neuroscientists were working to untangle the biological basis of intelligence. In the 1980s, a young medical doctor named Wolfram Schultz established his research lab focusing on Parkinson's disease. He intended to record the electrical activity of dopamine neurons to better understand their role in movement. Though dopamine neurons comprise less than 1 percent of all neurons in the brain, they're relatively easy to find and record from because they're concentrated in a cluster of midbrain areas. Schultz and his colleagues implanted macaque monkeys with electrodes and measured the activity of dopamine neurons while the animals performed simple motor tasks. Based on their known involvement in movement, the dopamine neurons might be expected to fire whenever the animals moved. Instead, they fired when the animals were given a reward.

This wasn't entirely unexpected. Though dopamine was known to be involved in movement, what is movement ultimately for, save the pursuit of reward and avoidance of punishment? Biologists had been slowly piecing together the basis of this orienting system. Nineteenth-century scientists, smitten with the newly discovered phenomenon of electricity and dimly aware of its involvement in the nervous system, stuck stimulating electrodes in the brains of humans and animals, guided more by exuberance than reason. By the early twentieth century, the neurosurgeon Wilder Penfield had refined this technique and used it to build an atlas of brain functions. He annotated a map of the brain with the effects of stimulation: the visual cortex was labeled "lights and shadows." Stimulation of a region he called "memory" caused patients to become immersed in flashbacks so vivid that it seemed as if they were playing out in the present. The frontal cortex, involved with executive functions, was termed "silence" because it extinguished patients' internal monologues. "Throughout my career," Penfield writes, "I was driven by the central question that has obsessed both scientists and philosophers for hundreds of years. Are mind and body one?" It appeared—almost too tidily—as though mental functions could be directly mapped onto distinct brain regions.

A flurry of stimulation experiments ensued, linking brain areas to their evoked behaviors. Stimulating deep in the brains of rats caused them to become aggressive. Stimulation of another area made them fearful, and these rats avoided returning to the places where they'd been zapped. In 1953, a postdoc named James Olds made a fortuitous discovery while performing his first electrode placement surgery in a rat. His surgery had been a fraction of a millimeter off target. The electrode impulse didn't make the rat fearful; rather, it seemed to be rewarding. The rat repeatedly visited the area where it had been zapped, instead of avoiding it. Like in a game of hot and cold, the animal could be "pulled" to any spot by delivering an electrical stimulus after each move in the right direction.

Olds quickly rigged up a lever system that allowed the animal to self-administer stimulation. The electrode-implanted rat pressed the lever incessantly. Later studies revealed that rats would choose to stimulate this area rather than eat, drink, or mate. They'd press the lever even if it was paired with a painful electric shock. Like addicts, animals would press the lever all day, every day, often until collapse, convulsion, or sometimes death. Rats, cats, monkeys, and even a dolphin died this way. These results captivated popular media, and futurists predicted that all striving and desire would soon be replaced with electrophysiological contentment. Science-fiction writer Isaac Asimov concludes: "Evidently all the desirable things in life are desirable only insofar as they stimulate the pleasure center. To stimulate it directly makes all else unnecessary."

Stimulating this area, it was later discovered, elicited dopamine release. Dopamine's reputation as the "movement molecule" was soon eclipsed by its new identity as the "pleasure molecule." Yet this was always a dubious association. Humans implanted with electrodes in an analogous brain region didn't experience pleasure when that area was zapped. It was more like the feeling of *wanting*. Subjects reported feeling compelled to press the lever, as though scratching an itch. Absent the ability to talk to animals, we can't say whether what we call reward feels pleasurable to them. Rather, it reinforces the behaviors that lead to it, as with Thorndike's law of effect. Dopamine isn't a measure of pleasure. It's more like the "motivation molecule," driving subjects to take actions to get what they want.

Schultz's new data complicated this picture. His team had confirmed that dopamine neurons fired when the animals received a reward—but only when they didn't expect it. The scientists had trained the monkeys to tap a lever to get a juice reward after a light cue was flashed. The untrained animals behaved, at first, randomly. They pressed the lever erratically, and this behavior was occasionally reinforced with juice when they happened to push at the right time. The dopamine neurons of these untrained animals fired when the reward was delivered. But once the animals were experts and had made an association between cue and reward, the dopamine neurons no longer fired in response to the juice delivery. Rather, they fired in response to the light cue that preceded it. Even more tellingly, if the light came on but the reward was withheld, the dopamine neurons reduced their firing. They were signaling what they expected to happen. The dopamine neurons weren't tracking movement, exactly, or reward; they were tracking *belief.* They even signaled when something they expected to happen didn't happen, as though surprised.

Psychologists already knew that surprise was an important part of learning. Animals don't always learn from simple repetition and reward. Surprise indicates that something remains to be learned—its attention-grabbing effect potentiates learning. Even infants pay more attention to surprising stimuli, such as a video of a ball rolling uphill. Surprise happens when experience violates expectations, and the brain uses this as a learning signal. This is why language programs often teach humorous sentences with unexpected associations: "Why is the banana wet?" and "My horses collect teeth." When teachers employ surprise, students are more likely to remember the lesson.

Schultz's findings also resembled classic results in psychology, like Pavlov's dogs salivating in anticipation at the sound of a bell signaling that food was soon to come. Decades later, the psychologist B. F. Skinner continued this line of research and dubbed behavioral feedback "reinforcement." Negative reinforcers, like an air puff, could abolish certain behaviors, while positive reinforcers, like food, strengthened them. He believed that this simple valence, the push and pull of desire and avoidance, was the axis along which all intelligent behavior could be built. In this way, he hoped to reduce all the complexity of animal behavior to a simple sort of physics, shaped by the attraction and repulsion of reward and punishment. Skinner believed that animals could be effectively "programmed" with reinforcers to behave in arbitrary ways. He thought that this was as true for humans as it was for any animal: "The real question," he writes, "is not whether machines think but whether men do. The mystery which surrounds a thinking machine already surrounds a thinking man."

Many mid-twentieth-century researchers believed that these findings would pave the way to a utopian and rationally designed "psycho-civilized" society. Fear and aggression would be tamed and pleasure amplified. Human behaviors would be logically steered. "It is a mistake to suppose that the whole issue is how to free man," Skinner argues. "The issue is to improve the way in which he is controlled." Inspired by this vision, the psychiatrist Robert Galbraith Heath tried to shape the behaviors of human patients using methods that were wildly unethical, even by the standards of his day. In 1972, Heath announced that he had successfully "converted" a gay man to heterosexuality by stimulating dopamine neurons while compelling his subject to have intercourse with a female sex worker. Skinner's ideas continued to dominate the field of psychology for decades. Humans were treated as passive objects, animated by punishment and reward and best controlled through nudges. Skinner's philosophy later heavily influenced the field of economics, whose practitioners call these reinforcers "incentives."

- - - - -

In the 1970s, there were only a few universities in the US where students could get hands-on experience with computers, and Stanford was one of them. Richard Sutton was an undergrad at the time. There was no AI major per se, so he majored in psychology and learned how to program on the side. Sutton was surprised by how little attention AI researchers paid to psychology. Surely AI should take inspiration from real brains; "Surely someone else would have studied this thing that animals and people are doing, a very commonsense thing." At the time, most game-playing AI systems had nothing to do with the way human intelligence works. They were based on brittle equations overspecialized to their tasks, vulnerable as panda bears adapted to bamboo alone.

In his junior year of college, Sutton came across an obscure and speculative tech report by a US Air Force researcher named A. Harry Klopf: *Brain Function and Adaptive Systems: A Heterostatic Theory.* Whereas a homeostat, like the thermostat of a house, is designed to maintain some status quo, a heterostat is a system designed to maximize something. Klopf believed that we should think of neurons (and organisms, neighborhoods, and societies, for that matter) as wanting something: having goals and acting in ways that maximize their future reward. While not all of Klopf's ideas have withstood scrutiny over time, his main argument was that learning is fundamentally hedonic.

Klopf's report made a profoundly clarifying distinction: intelligence can't be emulated by a passive program that chirps yes or no to classify pictures of cats or cars. Intelligence moves through, and acts on, the world in pursuit of goals, straddling a vector pointing away from punishment and toward reward. Intelligence *wants*. Sutton decided to pursue his PhD in the lab of AI researcher Andrew Barto, who shared his goal of building systems that learn the way living things do. Inspired by Klopf's insights, they spent the next few years formalizing Samuel's self-playing checkers program into an elegant framework now known as reinforcement learning. Its principle is simple: behaviors that lead to a reward state are reinforced, and those that do not are dropped. Echoing the experiments of psychologists past, Sutton and Barto pioneered the design of systems that learn, through trial and error, to express behaviors that maximize reward. Board games were a perfect medium for this, coming, as they do, with an inbuilt goal of winning.

A primary challenge for reinforcement systems is known as the credit assignment problem. Rewards come rarely in life. One could code up a checkers-playing program that "wants" to win. But a game's outcome is known only at its end, after a long chain of actions. How can a program know if all the moves that lead to a win are equally good, or if there is a deciding move that should get more credit than others? It's unclear how to assign such infrequent feedback to actions taken many steps in the past. Wins and losses are simply too rare to be reliable training signals.

To solve this, Sutton and Barto devised a training signal that can be updated at every turn of a game: the algorithm's prediction of whether it will win. A system tasked with achieving a goal must develop good predictions about how its choices lead toward or away from that goal. Sutton and Barto devised an algorithm that makes predictions about the outcomes of its choices and then compares these predictions to reality in order to improve them. They deemed this temporal difference learning. On a given turn of checkers, for instance, the program looks through all possible legal moves and estimates each move's likeliness of leading to a win. Say that it chooses an action it believes to be 90 percent likely to lead to victory. After a few more turns, the program is worse off than expected. It now estimates the likelihood of a win to be only 50 percent. Somewhere along the line, its prediction was wrong. This signal, called the reward prediction error, acts like surprise. It signals that something remains to be learned, causing the system to tweak its original guess to better align with reality. Sutton calls this "learning a guess from a guess." Eventually, the system will verify the accuracy of its predictions when it wins or loses the game. It plays against itself repeatedly, continuously improving its guesses to make choices that reliably lead to a win.

In the early 1990s, reinforcement learning scored a significant victory when the engineer Gerald Tesauro trained a temporal difference learning program to play backgammon. Though it lacks the cultural cachet of chess, backgammon had been immensely difficult for computer programs to crack. By 1990, chess- and checkers-playing programs were getting close to playing at the level of top human players. These wins, however, came largely thanks to brute-force programs that relied on massive computing power, rolling out all possible board configurations dozens of moves in the future. Some games are more amenable to brute-force methods than others. In checkers, there are, on average, three possible legal moves for a given turn. This is called the game's branching factor. Looking ahead quickly snowballs into a massive calculation, as each of those three moves generates three more possibilities, and so on. In chess, the branching factor is a more unwieldy thirty-five. Backgammon, a favorite pastime of humans for at least five thousand years, is a game of luck and strategy. Two players starting at opposite ends of the board must move their fifteen checkers across the board according to dice rolls. Backgammon's branching factor is around four hundred. Looking ahead, even for just a few steps, quickly becomes unmanageable.

Tesauro named his program TD-Gammon, after temporal difference learning. As Samuel had with his checkers-playing program, Tesauro pitted the program against copies of itself and let it amass a tremendous amount of experience through self-play. Like MENACE's, TD-Gammon's first games were terrible, and it won only by accident. It hadn't yet learned anything, so its predictions were pure guesses. After it played a few dozen games against itself, its moves came to resemble basic strategies of standard play. After three hundred thousand games, it played at the level of the best previous backgammon programs. This was particularly impressive because TD-Gammon hadn't been explicitly given any game expertise. Other top backgammon programs had been trained to copy expert human strategies and were programmed with intricate equations of game rules and dependencies.

TD-Gammon built up its systems of beliefs through experience alone. After millions of games, TD-Gammon held its own against top human players. Analysts who studied its play style found it had developed unorthodox strategies that challenged traditional wisdom. Tesauro noted that its unusual style led "in some cases to major revisions in the positional thinking of top human players." TD-Gammon demonstrated, for instance, an opening strategy superior to "slotting," then an almost ubiquitous early game play. Slotting has since all but disappeared from backgammon tournaments. Turing's dream had been realized: a program that taught itself how to play backgammon revealed—to experts, no less—a deeper understanding of this ancient game.

TD-Gammon, inspired by psychology, learned to play backgammon better than most people could. But was it learning the way humans learn? Theoretical neuroscientists Peter Dayan, Read Montague, and Terry Sejnowski made an unexpected connection: the strange firing patterns of dopamine neurons recorded by Schultz's lab looked exactly like the prediction error signal used in systems like TD-Gammon. Sejnowski reached out to Schultz, and the team came to collaborate on a set of now canonical papers suggesting that dopamine serves to broadcast the brain's reward prediction error.

This elegant framework explained Schultz's initially confusing findings. Though dopamine serves many functions in the brain, this was evidence that it acts as part of a learning algorithm. Like the temporal difference algorithm, the dopamine system encodes an animal's reward predictions. The activity of dopamine neurons reflects whether an animal receives more or less of a reward than it expected. It's rare, in neuroscience, to find so clear a marriage between theoretical model and experimental data. The brain formulates some internal world model and registers deviations from its predictions. As neuroscientist Robert Sapolsky puts it, dopamine is "about the pursuit of happiness rather than happiness itself."

Dopamine plays a crucial role in intelligence. It's responsible for reward-seeking behavior in most motile animals, and it is enriched in more intelligent species, like primates and humans. However, caveats are in order. What I've presented is already a vast oversimplification of what we know about dopamine neurons, and a fraction of what remains to be discovered about them. Dopamine neurons do not all signal the same thing. Some track the size and value of predicted rewards, but others appear to follow the animal's movements, motivation, punishment, uncertainty, or sensory predictions. Also, dopamine release impacts downstream neurons in myriad ways. It's a mistake to talk about dopamine as if it represents only one signal—an error I will continue to commit for the sake of simplicity. Reinforcement learning algorithms can take many forms, and we don't know for sure which version—if any—is coded in real brains. And, of course, there is much more to the story than dopamine. The dopamine system is only one component in a massively tangled learning network involving diverse populations of neurons, chemicals, and receptors throughout the entire organism. Ouroboric loops muddy the meaning of every neural signal. Despite these many caveats, reinforcement learning has served as a productive framework for beginning to explore one of dopamine's potential roles in the brain.

In this interpretation, planning can be thought of as the brain running reinforcement learning on its imagined experience, stepping through its internal model of the world. "Learning and planning," Sutton writes, "are fundamentally the same process, operating in the one case on real experience, and in the other on simulated experience from a predictive model of the world." Likewise, this might also explain the phenomena of regret, which can be thought of as a form of learning. Memories are data the brain can train on again and again. Rumination is the brain linking mistakes made in the past to negative outcomes experienced much later. Though commonly thought of as an emotion, regret might be better framed as a computational principle. This process can become pathological, however, as in the case of people with PTSD.

This framework also explains dopamine's diverse roles in psychological diseases. Sarpagandha, which lowers dopamine levels, was used for centuries by Ayurvedic doctors to treat schizophrenia. Later clinical research confirmed that dopamine-suppressing drugs dampen hallucinations. Like so many neurological drugs, these became standards of care before anyone knew why, exactly, they worked. But the reinforcement learning model of dopamine gives us a clue. Perhaps hallucinations result from dopamine's prediction system gone haywire: the brain's internal world model grown so strong that it drowns out reality.

Contrast the adult brain with that of a baby. Babies, argues psychologist Alison Gopnik, have little lived experience, so they're still forming their internal world models. Everything seems like a prediction error because they have no predictions to make. Nearly all their experience is pure surprise. As such, babies are motivated to explore everything. Gopnik calls this the "lantern consciousness" of children—the light of their interest lands on all around them. Adults, on the other hand, have a "spotlight" of attention, narrowly focused on whatever task is at hand.

This model of dopamine function has several important implications for biological intelligence. The first implication is that the dopamine system supports reprogrammable values. An arbitrary token predictive of reward, like a bell signaling food, becomes rewarding itself. This token can be anything. For laboratory animals, it might take the form of a light flash that indicates a sip of juice is coming. For humans, it might look like a favorable score on a test important for college admissions, a falling temperature indicating that one's child is recovering from their fever, a high game score, or a piece of colored paper meant to represent the abstraction we call money. It is, in short, how our economy works, how religion works, and how these abstract systems ever came to be.

Games are a perfect example of this. Humans can, for the purpose of play, slip in and out of collectively hallucinated value systems: harvesting stars in *Super Mario Brothers*, weapons in *Skyrim*, or craft recipes in *Animal Crossing*. Monopoly money is a classic case of something with illusory value. But for siblings absorbed in the heat of play, those paper slips are anything but valueless—at least until the game's end erodes all their power. This apparent mundanity reveals something profound about the human mind: the brain can learn to find ideas rewarding. And, though imagined, these currencies are still very real. People routinely backstab family members and friends in their temporary lust for Monopoly money. This desire dissolves, in an instant, with the game's conclusion. Nor is there a hard line between "real" (fiat) and "imaginary" (in-game) currencies. Gamers and gamblers take on real debts. Lawyers have noted a steady increase in divorce filings citing gaming debts. In 2022, a Russian soldier captured in Ukraine admitted that he'd joined the army to pay off debts he'd incurred in the *World of Tanks* video game series. Through social agreement, a seemingly worthless idea or token becomes imbued with value. The flexible nature of what we value, and learn to value, challenges assumption at the heart of modern economics—a tension we'll continue to explore throughout this book.

The dopamine system can also link actions to outcomes over impressively long timescales. Humans wield a prodigious, bordering on pathological, ability to work toward long-term goals—a talent made possible in part thanks to dopamine. In Schultz's original experiment, the dopamine neurons of trained animals came to fire in response to a visual cue that preceded the juice reward. When experimenters added an even earlier light cue, the dopamine neurons came to respond to the earlier light flash—albeit less strongly. But the brain can daisy-chain cues predictive of reward over long stretches of time. These cues may motivate, over many steps, animals' efforts toward long-term goals. The brain chains predictive signals along the way to its objective—the glint of a rice seedling, fresh deer scat—maintaining the motivation necessary to train, study, track prey, farm, or work on projects whose benefits will only be realized in the future.

Reinforcement learning algorithms rely on trial and error and thus require lots of experience, or training data. Customs like play provide people with social experience in a safe domain, allowing them to explore realms of possibility without taking existential risks. Samuel and Tesauro educated their child machines by pitting them against copies of themselves, generating massive amounts of simulated experience. But humans don't need to play millions of games to improve; they can learn from one another. They can simulate experience through imagination, play, and language. Language is a tool people use to modify their beliefs and reward expectations, enabling them to link cues over even longer timescales. We share experiences and collectively build beliefs about given actions by telling stories. Farmers don't each have to go through repeated trial and error to grow corn, because they can rely on the accumulated knowledge of their predecessors. Language replaces direct experience. It can be used to synthesize previous experience and install in other minds, at scale, beliefs that inform appropriate choices.

Beliefs make up the brain's model of the world, built through both personal experience and learning from others. Beliefs are reinforced not for being accurate but for leading to reward. An accurate belief *may* be reinforced—a hunter might use their correct knowledge of wildlife habits to feed their family, for instance. But beliefs can also be reinforced by social rewards, as with the propagation of conspiracy theories in insular online communities. It doesn't matter if these beliefs are accurate—they can be reinforced by social inclusion and the gamified social network feedback of "likes."

Humans are unique among animals in that they will even work for rewards they won't see in their own lifetimes. If they hold a shared belief strongly enough and the expected reward is large enough, humans can be motivated to behave over arbitrarily long timescales. Children will be on their best behavior for months in anticipation of Santa's annual judgment. Religious devotees commit to virtuous precepts, adhering to principles that they believe will win them entrance to heaven. This social phenomenon—belief in something one has been told about but doesn't have direct knowledge of—became reified as the virtue of faith. Humans learned to exploit their inbuilt reward systems to promote prosocial beliefs and behaviors.

Another important implication of the dopamine signal is that it collapses rewards into a single measure. Diverse rewards—food, sex, money, social belonging, drugs—elicit similar responses in dopamine neurons. For centuries, economists have employed the concept of utility—a scalar measure of a person's enjoyment of a good or service. Some scientists argue that dopamine is the biological reflection of that idea. Dopamine release is self-reinforcing; it etches in behaviors that are more likely to lead to more of its release. Addictive drugs like cocaine short-circuit this learning system. These drugs reinforce drug-seeking behavior by increasing dopamine levels. Drug seeking comes to replace formerly rewarding activities, such as eating or social bonding. The drug becomes an addict's principal source of satisfaction. Everyday environmental cues that are strongly associated with the drug can trigger the addictive behavior. Technology companies exploit this same principle to enthrall users to their platforms, using reinforcement mechanisms like novelty and social attention.

Interestingly, learning itself is rewarding to the brain. Researchers

have found that the "Aha!" moment of insight in solving a puzzle triggers dopamine release the same way sugar or money can. The brain
quite literally rewards itself for learning to make good predictions.
Game designer Jesse Schell reasons why this might be: "People who
enjoy solving problems are going to solve more problems and probably
get better at solving problems and be more likely to survive." People
love games because they offer free pleasure—an insight the Lydians
discovered in a time of famine. Rule-based games like chess are inherently predictable systems, orderly realms in which the brain can generate reward out of thin air, leveraging the gratification of making
predictions. Games are the reward system tickling itself.

This also explains why games can be so addictive. Muhammad al-
Amin, caliph of Baghdad, was rumored to have lost his kingdom because he was so engrossed in a game of chess that he failed to respond
to his brother's siege of the capital. King Canute of Denmark reportedly had Earl Ulf assassinated over an argument about chess. Today,
people still die during gameplay, and a handful of children have died
from neglect by their game-addicted parents. Since 2004, young men
have disproportionately dropped out of the US workforce, a decline
that researchers suggest is partly due to the availability of highly engaging video games. Consumers spend more money on games than on
other forms of entertainment combined. Problematic gamblers make
up around 1 percent of the global population, and about 3 percent of
the global population are problematic gamers (people who have un-
controllable urges to play video games). For comparison, alcoholism
affects about 1.4 percent of people worldwide and opioid addiction
about 0.2 percent globally.

There are two ways the brain can improve its predictions. The first
is to build better—more accurate, more nuanced—models of the
world. This is the aim of science, scholarship, and simulation. The
second is to render the world more predictable. This is the realm of

technology and niche construction, and the genesis of our procrustean impulse to shape the world to fit our beliefs about it. We humans domesticated plants and animals to ensure a stable food supply. We invented housing and clothing to protect against the elements. We domesticated ourselves as well: by adhering to social norms dictated by culture and religion, we have rendered one another more predictable. Like game players bound by the rules of the game, humans are bound by social convention, by the beliefs or world model according to which we collectively agree to behave.

Games, long dismissed as trivial—a domain for children and drunks, cheaters and braggarts—reflect a deep operational principle of the brain. An algorithm designed to learn checkers and backgammon illuminated the role of dopamine, an evolutionarily ancient learning system. It's an algorithm that's responsible for many of our cultural innovations, from money to games, science to religion. It's little wonder that games have become a dominant metaphor through which our understandings of chance, economics, evolution, society, epistemology, intelligence, and war have come to be colored. They've been such a productive metaphor because they are an outgrowth of what minds do. Brains are prediction engines, and nowhere has that been more evident than in our fascination with games of chance.

3

Dice Playing God

All chance, direction, which thou canst not see.

ALEXANDER POPE

hance was once believed to relay the will of God, to be the means through which the sacred plane pierced into the mundane world. Humans attempted to contact that power through augury rituals, intuitively linking the unknowability of dice rolls with the unknowability of the future. Working in the mid-twentieth century, Omar Khayyam Moore was a Yale anthropologist interested in understanding how people solve problems. In particular, he hoped to find strategies universal across cultures. In the 1950s, he grew puzzled by a widespread human practice lacking obvious efficacy: divination. "Magic is, by definition and reputation, a notoriously ineffective method for attaining the specific ends its practitioners hope to achieve through its use," he writes in a 1957 paper. Divination features prominently in most cultures known to history, so why would humans across millennia cling to magic if it never worked? Were these traditions really so ineffective?

Moore forwarded another interpretation. Previous researchers had assumed that the practice of magic couldn't possibly lead to its desired results. What if magic *did* work—just not in the way it purported to? He refers to practices of the Innu people of the Labrador Peninsula. "Theirs," writes anthropologist Frank Speck, "is almost wholly a religion of divination." They regularly consulted the gods on pragmatic, existential matters, like determining the direction in which they should hunt for game. The gods whispered their promises through the cracks of roasted animal bones. To fashion the mouthpiece of a god, the shoulder blade or hip bone of a fresh kill was pared of all meat, boiled clean, hung to dry, and fitted with a wooden handle. The shoulder blades of caribou were known for being particularly truthful. The prepared scapulae were held over hot coals in a divinatory ritual, causing cracks to radiate across the bones' smooth planes. These cracks, in conjunction with the consultant's dreams, indicated which way hunters should go for their next pursuit, pinning their vision to a cardinal direction. This method was reserved for times of uncertainty, when the consultant had no reliable information about where their prey might be.

Of course, there aren't control groups in anthropology, but Moore invites his readers to conduct a thought experiment. Imagine that the Innu had undertaken the same ritual without using the bone cracks to determine a direction. What would influence the direction they chose for their hunt? Personal preferences, likely biased by previous successes. Because heat-induced bone cracks and the animal's location were unrelated, divination was a randomization technology. The method helped the Innu avoid fixed patterns in their hunting strategies, rendering them less predictable to prey. Moore notes that mathematicians had recently proved that randomness is the optimal strategy for certain games. In the game Rochambeau, for instance, the best possible strategy is to throw rock, paper, or scissors randomly, with equal probability.

Using randomness to confound one's opponent is a brilliant strategy, and one much older than human history. The nervous system of
many moths, for example, has a switch that's flipped when it detects
the echolocating sweep of a bat. The moth's flight path then turns into
a chaotic tumble, rendering the insect less predictable to animals
pursuing it. Many mammals use a similar tactic. Likewise, birds aren't
always able to fly faster than their predators, but they can often out-
maneuver them by introducing unpredictable elements into their flight
patterns.

Randomness was one of humanity's earliest cognitive aids. Unlike
moths with a nervous system switch for evading bats, people don't
have an intrinsic way to produce randomness—they tend to fall into
patterns. But humans across the globe discovered methods for generating chance events with instruments like bones, cards, and dice. The
word *forecast* originally referred to divination of the future (including
weather) by casting lots. The Dogon read the footprints of foxes they
lure to the edges of their villages with food left out at night. The Dalai
Lama, the leader of Tibet, has been known to draw names from a bowl
to reveal his successor.

The Greeks regarded divination as a revealed technology. They
credited the mythical figure Prometheus with gifting it to humans,
alongside fire and metallurgy. Religious leaders consulted polyhedral
dice inscribed with the names of various gods to inform supplicants
whose benediction they should seek. The Bible depicts Israelites drawing lots to make important decisions, ranging from military drafts to
distributing Israel's territories among tribes. As the book of Proverbs
has it, "The lot puts an end to quarrels." Martin Luther described the
casting of lots as "a real act of faith," arguing that it accurately reflects
God's decisions. In the eighteenth century, a religious sect known as
the Moravian Brethren eschewed reliance on human reason—then

much in fashion—and instead adopted lots to dictate all its decisions, believing that this represented the true will of Christ. In this manner, lots decided the layout of present-day Winston-Salem, set up marriages, and settled questions of morality for the Brethren.

In countless cultures across the globe, gaming devices doubled as divination and decision-making tools, from the sacred altars of the Hopi to the ritual spaces of ancient Levant. Although the practice of divination was widespread, most religions restricted who was allowed to interpret its results. The Bible prohibited regular people from using lots to make decisions, though representatives of God could. Perhaps chief among the decisions made in this way was the apportionment of land by lottery—for example, dividing estates fairly among heirs. It was so fundamental to the priestly profession that the Greek word *kleros*, meaning "lot, chance," gave rise to the word *clergy*. God may not play dice, as Albert Einstein famously asserted, but sometimes dice play God.

More than a decision-making aid, randomness also features in wildly popular games of chance. Many cultures throughout history have had some form of gambling, which brought some societies to the brink of ruin. Knucklebone dice, or astragali, have been found among the remains of seven-thousand-year-old human settlements in present-day Iran. To be fair, it isn't always possible to definitively identify game pieces unearthed at historical sites—the identification of dice has been likened to an archaeological Rorschach test. But we do know that by 3000 BCE, gambling had become such a problem in Egypt that it was outlawed. Millennia-old Hindu legends recount rajas gambling for days on end, some playing unto ruin, exile, or slavery. Gambling, they recognized, was like a drug: "As alluring as a draught of Soma on the mountain / the lively dice have captured my heart."

The flow of valuable goods across pre-Columbian America appears to have been driven by a continent-spanning gambling network. John

Ordway was a member of the 1804–1806 Lewis and Clark Expedition that mapped the territories west of the Mississippi River. He was surprised to discover members of the Nez Perce Tribe in present-day Idaho gambling with iron axe heads that had been traded to Indigenous Americans the previous year in present-day North Dakota, over a thousand miles away. Indigenous American women were renowned for their prowess in dice games, and many appear to have been buried with their hard-earned winnings. Gambling largely happened between, rather than within, tribes and thus served to drive goods across the continent. Betting against a fellow tribe member was less acceptable, as it was said to be like "winning from yourself." Gambling found success as a technology for reallocating resources across cultures, classes, and space—similar to the role of markets in Eurasia.

Leaders throughout history have repeatedly banned gambling. Today, it's most notably prohibited among practicing Muslims, as the Quran expressly forbids it. In some cultures, however, it achieved widespread acceptance. The ancient Greeks and Romans delighted in games of chance; Socrates describes a festive palaestra packed with young men playing dice. Many Roman emperors were profligate gamblers. According to nineteenth-century historian Andrew Steinmetz, Augustus was pathological, though generous. Caligula was caught lying at the table, while Nero gamed with the intensity of a "madman" and cheated with weighted coins. Claudius, who played "like an imbecile," had the interiors of his carriages fitted with game boards and authored a thoroughly incorrect book on how to win at dice. In his excavations of ancient Rome, the archaeologist Rodolfo Lanciani found makeshift game boards engraved onto "almost any flat surface" accessible to the public.

Reflecting on the last days of the Roman Empire, Steinmetz claims: "Finally, at the epoch when Constantine abandoned Rome never to return, every inhabitant of that city, down to the populace, was addicted to gambling." Gambling's thrall over Europe never dimmed. Historians write of men wagering their wives and children. They bet their fingers, eyebrows, and freedom, offering themselves as slaves to the winners. Some even wagered their lives. The seventeenth-century French legal scholar Jean Barbeyrac cites a gambler who willed his skin and membranes to cover a gaming table, and his bones to be made into dice. Barbeyrac laments gambling's grip on his countrymen:

> I do not know if there is any other passion which allows less of repose and which one has so much difficulty in reducing. [. . .] But the passion of gambling gives no time for breathing; it is an enemy which gives neither quarter nor truce; it is a persecutor, furious and indefatigable. The more one plays, the more one wishes to play; one never leaves it, and with difficulty, one resolves to leave off a little while from dice and cards to satisfy the needs of nature; all the time he is not playing, the time seems to him lost; he is tired (ennui). When he does anything else, it seems that gambling had acquired the right to occupy all his thoughts.

The Renaissance was so named for marking a "rebirth" of reason; it was a time when scholarship and logic were celebrated as the crowning achievements of society. At the same time, the cultural obsession with gambling threatened to depose the top echelons of society. Many Christian cultures considered dice immoral, given that the Bible depicts Roman soldiers using dice to divvy up Jesus's clothes while he died on the Cross. But gambling was given a pass during Carnival season, a time when lawmakers overlooked normally verboten behaviors. In Venice, the birthplace of professional casinos, Carnival came to take up six months of the year. Members of Venetian nobility were already seeing their fortunes shrink due to the plague and related trading losses, and gambling accelerated their bankruptcy. Members of the aristocracy lost fortunes amassed over generations in the span of a few hours, perhaps all the easier for not having been hard earned. The city eventually voted to outlaw gambling entirely.

By then, however, casinos had spread across Europe. German spas, touting their healing waters as panaceas, doubled as gambling dens. Bored with the rest prescribed by their doctors, wealthy patrons sought entertainment during their health retreats. Gambling fit the bill. The Russian writer Fyodor Dostoevsky ruined himself financially several times over in German spas, an experience that eventually inspired his novel *The Gambler*. In France, gambling—though technically illegal—was a popular pastime among all levels of society, and it was effectively required of the ruling classes. Jules Cardinal Mazarin, chief minister to Louis XIV, was accused of pushing gambling on the upper strata of society to keep the nobility in thrall, their minds happily occupied by matters other than subterfuge. Their growing gambling losses also kept them indebted to the king. "Games of chance came into vogue," writes a contemporary, "to the ruin of many considerable families: this was likewise very destructive to health, for besides the various violent passions it excited, whole nights were spent at this execrable amusement." The egregious outlay of state funds on gambling and frivolity was one of many sparks that ignited the French Revolution.

Gambling was a problem not only for the elite, however. The eighteenth-century politician Jean-Joseph Dusaulx claims to have "found cards and dice in many places where people were in want of bread." Farmers gambled away their entire harvests, he reports, and merchants staked their wares. The result of this state of things, writes Steinmetz, was incalculable social affliction. "At the death of Louis XIV," he asserts, "three-fourths of the nation thought of nothing but gambling."

To bring the situation under control and profit from the frenzy,

Napoleon legalized France's main gambling clubs and, by taxing them, tapped an enormous source of revenue for the state. Humanity's endless fascination with uncertainty would become a source of inexhaustible tax income. As private and public institutions drew massive profits from gamblers, it was only a matter of time before mathematicians began treating games as a serious field of study.

Given that gambling predates written history, many scholars have wondered why the mathematics of probability was discovered only relatively recently. People had long understood what constituted the space of all outcomes and what might be considered probable or improbable within that. Several ancient traditions had the ingredients for calculating a probability: Greek and Chinese divination manuals enumerated all possible outcomes for casting multiple dice or yarrow stalks, linking these to fortunes. Why, then, did a mathematical treatment of probability not arrive until the Renaissance? One possibility, suggests physicist Shmuel Sambursky, is the long-standing conceptual gap between randomness and the idea of natural law. The ancients conceded, for example, that stars, so regular in their movements, obey fixed mathematical laws. However, they believed that terrestrial events are divorced from this heavenly causality and subject instead to the unknowable whims of capricious gods. No event seemed more unruly than a chance roll of dice. The Stoic philosopher Chrysippus characterizes dice rolls as events whose outcomes are caused "according to some hidden principle." In the Stoic worldview, the universe is deterministic, and chance is merely a form of ignorance, impenetrable to human intellect.

Sambursky cites another factor in the late discovery of probability theory: the ancients were content to philosophize but had less of an empirical—that is, experimental—tradition. It was not an accident that the rise of probability theory coincided with the rise of experimental science in the sixteenth and seventeenth centuries. It would take the data and mathematical insights of Kepler, Galileo, and Newton to reveal that the same gravity that governed celestial orbits also governed the trajectories of prosaic objects here on earth. As the statistician Florence Nightingale David puts it, though the ancients had plenty of practice with gambling, they also had "a habit of mind which made impossible the construction of theoretical hypotheses from empirical data."

David argues that the ancients were also hampered by their numeral systems. Greek and Roman numerals were clunky for algebra, though Roman numerals were well suited to the abacus calculations that dominated commerce and research for hundreds of years. What we now call Arabic numerals originated in India sometime in the sixth or seventh century. These were introduced to Europe through the writings of the Middle Eastern mathematicians al-Khwarizmi and al-Kindi around the twelfth century, though most European scholars didn't adopt the system until the fifteenth century. Arabic numerals were much better suited for representing fractions, essential to probability theory. They were also generally more intuitive: in Arabic numerals, for example, bigger numbers have more digits because the magnitude of a number is represented by its position within the number, with zeros as placeholders. In Roman numerals, even numbers that are very close to one another have very different numbers of digits. Take, for example, five (V), eight (VIII), and nine (IX). Algebraic symbolism had also improved with the introduction of plus, minus, and equal signs. In 1557, Robert Recorde introduced the first known use of the equal sign in his *Whetstone of Witte*. It was chosen, he writes, "because noe 2 thynges can be more equalle" than "a paire of paralleles."

Scholars made quick progress once they had the economic incentives and the mathematical symbolic systems necessary to represent chance. The earliest known treatment of probability was formulated in the sixteenth century by Italian scholar Gerolamo Cardano. The illegitimate child of a lawyer, he survived his mother's attempt to abort him and outlived his siblings, all of whom died of the plague. Cardano trained as a doctor but was denied a license to practice, owing to his illegitimacy and combative nature. Instead, he came to be known for his broad-ranging scholarship, including in mathematics and the natural sciences. Like many scientists of the era, such as Galileo, Cardano was made a celebrity by the horoscopes he drew up for the ruling class. Astrology played the role of today's science-funding agencies. Kings, merchants, and generals sought Cardano's advice on the most auspicious dates to begin military campaigns or strike trade agreements. He was also a celebrated philosopher: Shakespeare's "To be, or not to be" soliloquy in *Hamlet* was heavily influenced by a translation of Cardano's philosophical work *De consolatione*, widely believed to be the book that the titular character pores over in the scene. "Seeing, therefore, with such ease men die," writes Cardano, "what should we account of death to be resembled to anything better than sleep, &c. . . . Most assured it is that such sleeps be most sweet as be most sound."

Despite his many talents, Cardano was forever short of money; he found it impossible to turn out servants or deny his children pets, complaining that his house had grown littered with "kids, lambs, hares, rabbits, and storks." His obsessive scholarship, he claims, was a "counterpoise to an insane love for my children." Cardano confesses that he was, from youth, "immoderately given to gambling." He gambled his family into poverty on at least one occasion and admitted that his personal life suffered for his addiction.

In his book *Liber de ludo aleae* (*Book on Games of Chance*), likely finished in 1563 but only printed a century later, he analyzes throws of one, two, and three dice, as well as card games and effective cheating methods. This was the first time that anyone had analyzed a game's "odds": the ratio of favorable to unfavorable outcomes. Cardano determined that the probability of a die landing on any of its faces is equal, abstracting mathematical theory from empirical study. He also notes, without proof, that a measurement will become more accurate over repeated trials. One might roll a fair die ten times and, by chance, get a six every time, so the measured probability of rolling a six may initially look like 100 percent. But such luck can't hold out forever, and after three hundred rolls, the average value will be much closer to the "true" probability of approximately 16 percent. Jacob Bernoulli would later formalize this insight as the law of large numbers.

It's unclear why Cardano didn't publish *Liber de ludo aleae* in his lifetime; perhaps he thought it too trivial, or perhaps he thought better of revealing his cheating techniques and forfeiting the edge he needed to elude poverty. It would be another hundred years before scholars seriously took up the topic of probabilities again. By the seventeenth century, France had begun to eclipse Italy as the center of European scholarship. Blaise Pascal, born in 1623, was a sickly child prodigy with an overbearingly ambitious father. By nineteen, Pascal had invented a mechanical calculator and demonstrated the existence of a vacuum. In 1646, when Pascal was twenty-three, his father converted his family to Jansenism, a stringent religious movement within Catholicism. Pascal was prohibited from thinking about math, as Jansen held that scientific curiosity was just another form of sexual indulgence, akin to masturbation. When Pascal's father died in 1651, Blaise entered a brief "dissolute" period. He gambled at the behest of his doctors, who promoted it as a relaxing diversion that might cure him of his excessive mental exertion. During this time, Pascal was credited with inventing the roulette wheel, an unintended consequence of his attempt to develop a perpetual motion machine. More significantly, his new vice brought him into the orbit of the essayist and inveterate gambler Antoine Gombaud, who posed two questions to Pascal that brought about an entirely new field of mathematics.

Pascal lived in a time before academic journals existed, when

mathematicians staked their claims on solutions to outstanding problems through letters. In 1654, Pascal wrote to a more established mathematician, Pierre de Fermat, with the questions posed to him by
Gombaud and offered tentative solutions. After years of playing dice,
Gombaud claimed that if he bet on rolling at least one six in a series of
four throws of one die, it was on average a profitable bet. As a variation, he began betting on what he believed to be equivalent: that two
sixes would come up at least once in a series of twenty-four rolls of
two dice. Yet the second bet, he'd observed, was not profitable. Why
would the first win in the long term but the second lose? Pascal carefully calculated the odds and found that these two scenarios weren't,
in fact, mathematically equivalent. The likelihood of the first is a little
over one-half, whereas the likelihood of the second is a little under
one-half. With this calculation, Pascal defined the quantity we now
know as a probability.

Another question posed by Gombaud, known as the problem of
points, was first put forth by the Italian monk Luca Pacioli in 1494. He
imagined two players in a game of balla, which is played in rounds and
won once a player collects six points. In Pacioli's scenario, one player
has five points and the other three points when the game is interrupted
before they can finish. How should they split the winner's pot, assuming both players have equal skill? Pascal and Fermat enumerated all
possible outcomes to determine what share of the pot should go to
each player in a manner proportional to the likely outcomes. This was
the first known measurement of what would come to be known as a
game's expected value. It laid the groundwork for the mathematical
characterization of a fair division and the use of probabilistic arguments to reason about future returns.

These letters opened a new empirical dimension: here was a way to
measure the *future*. The notion of probabilis—a legal term that originated in Roman courts and referred to the credibility of a past event—

could now be extended forward. Probability theory gave rise to decision theory, the study of how rational agents make decisions (like how to split a winning pot). This would go on to become a cornerstone of economic thought. Pascal made this connection explicit, comparing gambling to a form of decision-making under uncertainty. Gambling was, in a way, a model of life: given limited information, one attempts to place wagers, or make decisions, that lead to the highest possible payoffs.

Take, for instance, the thought experiment now known as Pascal's wager. Humans, he posits, are faced with a fundamental uncertainty: whether God exists or not. This dictates how one should lead their life. There is no way of knowing, through reason, which of these possibilities is true. Nevertheless, humans must wager on one or the other, hazarding their lived experience. Now, suppose our player decides to wager that God exists. If this is true, the player stands to gain everything (eternal life in heaven). If it isn't, they stand to lose little, save the cost of virtue. Say, on the other hand, the player wagers that God doesn't exist. If this is true, the player gains nothing. If it's false, they stand to lose everything (an eternity of suffering in hell). One should wager, then, that God exists, even if the chance of that is vanishingly slight. This is the equivalent of staking a finite amount (one's behaviors during their lifetime) for the possibility of infinite gain. "Let us weigh," writes Pascal, "the gain and the loss in wagering that God is. [. . .] If you gain, you gain all; if you lose, you lose nothing. Wager, then, without hesitation that He is." Today, thinkers use similar arguments to reason about the risks of potentially world-altering developments, like catastrophic climate change. As the science of perspective allowed Renaissance artists to represent how people see, the science of games helped Renaissance thinkers represent how people decide—or, at least, how they believed people *should* decide.

Pascal and Fermat's 1654 exchange thrilled academia. It became clear that chance could, to some extent, be systematized, even domesticated. The following year, the Dutch polymath Christiaan Huygens traveled to France to study for his law degree, and a random introduction to the idea of probabilities in a salon led him to publish what would become a standard textbook on probability theory, *De ratiociniis in ludo aleae* (*On Reasoning in Games of Chance*). The book formalized the young field, focusing on topics including the problem of points and calculating expected values for a given event (for example, the expected value of the roll of a die is three and a half). Games, he argues, are deeper than they seem at first blush: "The reader will soon understand that I do not treat here a simple game of chance but that I have thrown out the elements of a new theory, both deep and interesting." He articulates the tension at the heart of probability theory, which had served to shield it from scholarly dissection for so long: "Although in games depending entirely on Fortune, the Success is always uncertain; yet it may be exactly determin'd at the same time, how much more likely one is to win than lose." This was a strange new kind of measure: what was inescapably unpredictable in a single instance—a given dice throw, or wheel spin—could be predictable in aggregate.

Scholars promised that probability theory would turn decision-making into a precise science. In his 1713 treatise, *The Art of Conjecturing*, Jacob Bernoulli contends that the field will have profound practical implications and that the art of precisely measuring probabilities will allow people to make choices with better, safer, and more advantageous outcomes. In the same year, the French mathematician Pierre Rémond de Montmort published his *Analysis of Games of Chance*, in which he extensively justifies working on the seemingly trivial field of games to better understand people's irrationality and superstitions. People, he argues, are quick to blame Fortuna for both good and bad outcomes. They try to appease her by following imagined rules, when they'd be better off blaming themselves for their poor decisions and learning the laws of chance to secure better fates. Although people had once used augury to eke out hazy guidance from a higher power, probabilistic reasoning would come to eclipse their reliance on divining God's will.

The investigation of an irrational activity, gambling, brought Enlightenment thinkers to a new concept of rationality. They believed that mathematics would correct the errors in human perception and undo the biases plaguing human thought. Take, for instance, the gambler's fallacy. Say that, in a series of coin flips, the coin has landed on heads ten times in a row. People often believe that the next toss is more likely to land on tails: the player is "due" for tails. But a coin's chance of landing on heads is *always* one-half, no matter what has come before.

Probability theory promised to serve as a cognitive aid. Instead of arguing with words, we'd argue with numbers. As the French mathematician Pierre-Simon Laplace would later put it, probability theory amounts to "common sense reduced to calculus." Probability, he writes, is "only the expression of our ignorance of the true causes." Our finite intellects limit what we can know, and therefore we're forced to traffic in possibilities and hypotheticals. It was the first hint that the mind itself might be in some way mathematical, knowable—that thinking could be cleansed of superstition and lifted into the Platonic realm of numbers.

Probability theory has helped excavate many human idiosyncrasies. One example, formulated by Nicolaus Bernoulli—Jacob's nephew—involved making a bet with an exponentially increasing payoff. A coin is tossed until it comes up heads, at which point the game ends. If the first throw is a head, the payoff is two ducats. If the second toss is a head, the payoff is four ducats. If the third comes up heads, the payoff is eight, and so on. If the nth throw is a head, the payoff is 2^n. The game's expected payoff is unbounded: one could conceivably continue getting tails and failing to get heads indefinitely, and therefore win infinite ducats. In reality, no one would wager a huge sum to play the game. Many mathematicians tried to reconcile this divide between common sense and calculation. One solution, elaborated by Nicolaus's cousin, Daniel Bernoulli, involved an idea that would become crucial to economic theory: marginal utility. A ducat given to a wealthy man benefits him less than it would a poor man. Past a certain point, infinite potential gain becomes effectively meaningless.

Other thinkers argued that probability theory was more than a lens to clarify human thinking—it was a model of thought itself. The Reverend Thomas Bayes was an amateur mathematician and a well-respected scholar. His most important work would only be published posthumously in 1761, after his friend Richard Price went through his papers and discovered an essay that would prove to be one of the most influential papers in statistics: "An Essay towards Solving a Problem in the Doctrine of Chances."

The idea at the heart of Bayes's theorem is simple. To infer truths about the world, one starts with a guess, informed by what one already knows (called the prior), and updates that guess given new evidence. It is an algorithm, not a static formula, to be iterated with every new piece of data. Say that someone smells smoke from their kitchen (evidence). They might worry that their house is on fire, but they quickly remember that their spouse said they'd be barbecuing for dinner (the prior). Given this prior knowledge, the more likely explanation is that the barbecue caused the smoke, not that their house is on fire. Bayes transformed rationality into something probabilistic in nature.

Bayes's theorem also clarified an important point: in the real world, the probability of an event often depends on many factors. The probability of landing tails on a coin flip is always one-half, but the probability of a person having a particular cancer often depends on their age, sex, and genetic background. Say that a blood marker becomes elevated in the case of both arthritis and a certain cancer. If the cancer is exceedingly rare, while arthritis is relatively common, it makes more sense, according to Bayes's rule, for a doctor to order follow-up tests for a potential arthritis diagnosis first. Given measured effects, Bayes's rule allows us to infer the most likely causes.

Laplace independently derived Bayes's results in 1774, imagining a man who had been living in a dark cave his whole life finally emerging into the outside world. He witnesses the sun rise and thinks that this ecstatic visual celebration is a one-off event. But with every day's new dawn, he grows increasingly confident that it's a common occurrence. Given repeated observations, our cave dweller updates his guess as to whether the sun rises every day and eventually comes to agree with E. E. Cummings's assessment: the sun is "never a moment ceasing to begin / the mystery of day for someone's eyes." This process is called induction, wherein one infers a general principle from repeated evidence. Today, Bayes's rule drives a host of functions in our everyday lives, from spam filters to the automated detection of medical and financial anomalies. In World War II, Alan Turing used Bayes's theorem to break the Enigma code securing German naval communications. Many neuroscientists argue that the brain performs Bayesian inference, filtering new information through its prior model of reality. Some philosophers of science suggest that the very practice of science is itself Bayesian.

Nevertheless, Bayes's theorem struck some scientists as clearly wrong. It seemed absurd that subjective belief should have any place in the realm of mathematics. In Bayes's model, it's not that reality isn't determinate; rather, we don't have direct access to it. We can only sample it through the scrim of uncertainty: the mysteries of dark matter, the indeterminacy of the quantum realm, the opacity of human behavior. All we have is our limited observations, and all we can do is update our beliefs according to the evidence we gather. The best we can hope for is induction: approaching truth in the limit and inferring general laws from particular instances.

Bayes's theorem is ubiquitous in the modern world, but due to its subjective nature, it can only tell us how to update our beliefs, not how to set our priors. Priors, therefore, are a place where biases can be neatly hidden. In deciding whether a new pesticide is safe, we can imagine two possible base assumptions. The first is that the product is potentially harmful to humans and the environment, and therefore requires a great deal of evidence to prove its safety. Medical experts and environmentalists favor this prior, called the "precautionary principle," as the safest stance for research and development. Applied as a broad-brush principle, however, it can be needlessly onerous and may prevent safe drugs from making it to patients who need them. An alternative prior deems a product safe by default and requires a lot of evidence to prove that its harms outweigh its benefits. Corporations and industry groups seeking more immediate profits naturally favor this stance, though it has resulted in serious disasters, from thalidomide-induced birth defects to environmental decimation by DDT. These philosophical stances affect our lives in many ways, from the adoption of drugs to major national public health decisions. But it's important to note that neither prior is more "data driven" or clear eyed than the other. Neither prior can serve as a one-size-fits-all blanket assumption.

Scientists began applying probability theory to questions of practical significance, from quantifying errors in scientific measurements to informing government policies. Though the earliest recorded documents in human history routinely enumerate wealth and economic activity, there seems to have been a long-standing taboo on enumerating health. Well until the late eighteenth century, some philosophers implied that to count the sick, or even the number of babies born, was tantamount to impiety: probing the very mind of God. The sixteenth century saw European countries, devastated by plague losses, taking their first steps toward amassing and centralizing big social datasets. In the 1750s, French scholars argued over whether it was rational to inoculate the general population against smallpox, which was then devastating Eurasia. Smallpox had a 30 percent mortality rate. Unlike the safer cowpox vaccines that would be developed in the 1790s by Edward Jenner, the early inoculant involved directly infecting patients with weakened smallpox. The practice had been successfully undertaken for decades in Turkey, China, and England, but it carried a nonnegligible risk: one in two hundred patients died from the procedure. When did it make sense to accept the small risk of immediate death to vastly reduce the probability of death by smallpox in the future? Daniel Bernoulli calculated the mathematical expectation and overwhelmingly found in favor of vaccination. Despite his careful reasoning, the French continued to eschew inoculation. The Italian physician Angelo Gatti recognized that individuals weigh personal risk not through impartial mathematical calculation but through subjective psychology: "An immediate risk, no matter how slight, will always make a greater impression than a very great, but distant and uncertain one."

In time, however, statistics would come to regularly inform policy. The nurse Florence Nightingale is famous for her use of data in pioneering many modern medical practices. She countered the philosophers who found it impudent to tally health data:

The true foundation of theology is to ascertain the character of God. It is by the aid of Statistics that law in the social sphere can be ascertained and codified, and certain aspects of the character of God thereby revealed. The study of statistics is thus a religious service.

The new field's success expanded to other social sciences. The nineteenth-century Belgian mathematician Adolphe Quetelet was a tireless advocate for a new science he called "social physics," whereby he hoped to work out a human correlate of celestial mechanics and describe the trajectories of civilization. Quetelet used massive datasets to detect underlying patterns. He found regularities in births, deaths, suicides, and even crime rates. He argued that crimes should not be considered acts of individual volition. Instead, he found that they obeyed statistical laws at the population level. "Society itself contains the germs of all the crimes committed. It is the social state, in some measure, that prepares these crimes, and the criminal is merely the instrument that executes them." Criminals were themselves victims, helpless to the laws of social physics acting on society, a necessary sacrifice to preserve civilization. He suggested that evidence-based legislation—aimed more at ameliorating the institutional causes of crime and less at punishing individual criminals themselves—was the best hope for reforming society. Quetelet's ideas were so pervasive that physicist James Clerk Maxwell often illustrated his work in statistical mechanics using analogies drawn from social physics. To those armed with the hammer of probability theory, the world looked like a statistics problem. If we could use math to understand even chance, why not use it to understand humanity itself?

At the same time, statistics gained a reputation for being a method of lying, more credibly, with numbers. As an idiom popularized by Mark Twain has it: "There are three kinds of lies: lies, damned lies, and statistics." Even practitioners with the best intentions fall prey to the nuances of its applications. This might be best illustrated with the problem of Gaussian noise. In 1809, the mathematician Carl Friedrich Gauss discovered that the error, or "noise," in astronomical observations follows a particular pattern, now known as a bell curve, or normal distribution. This pattern of noise—the same that governs the randomness of dice throws—seemed to be everywhere scientists looked: in the distribution of human height and weight, the motion of pollen grains suspended in water, the size of butterfly wings. It seemed so ubiquitous that many scientists used it to simplify the measurement of errors in a wide range of datasets, including many they shouldn't have. They were, as mathematician Nassim Nicholas Taleb puts it, "blinded by bell curves."

In 1900, almost one hundred years after Gauss's discovery, Louis Bachelier published an analysis of financial markets, modeling stock prices as a stochastic process with Gaussian noise. His early mathematical treatment of financial markets was adopted as the foundation of many twentieth-century economic theories. Bankers used the 1973 Black–Scholes model to price complicated financial derivatives based on factors including a stock's current price and volatility. Because volatility can't be measured in real time, only retrospectively, traders would guess its value by assuming that it followed a normal distribution. In the 1960s, the mathematician Benoit Mandelbrot warned economists that stock market volatility isn't Gaussian. Herd behavior induces crashes and bubbles, and outlier events happen more frequently than they do in Gaussian distributed data. This, Taleb writes, is the "ludic fallacy": the mistaken application of the well-behaved statistics of dice to describe the more violent randomness of the real world. As a result, the Black–Scholes model massively underestimated rare events. Traders using price predictions from the Black–Scholes model to offset risky gambles ironically invited more risk. They used the model to build an intricate derivatives market that collapsed like a house of cards in 2008. Related misapplications of statistics likely play a role in the ongoing replication crises in various fields of science.

Many things that we may think of as random are not necessarily so. For instance, a practiced dice thrower can reproduce a dice roll with the reliability of a professional athlete. A discussion of what it means to be genuinely random merits its own book, but perhaps the simplest definition is a sequence that cannot be predicted. Probability theory deals with the realm of things we do not, and often cannot, know. As such, it's inextricably linked with epistemology, or knowledge and how we know it. In the twentieth century, physicists discovered that certain quantum events can be truly uncaused and unknowable. Randomness isn't a by-product of human ignorance but an objective register of reality. The universe is fundamentally unpredictable. This can be profoundly frustrating and thoroughly irresistible to the brain, which, as we've discussed, is a prediction engine that seeks to ascertain even faint correlations and causal relationships. Our impulse to gamble, it turns out, is linked to the brain's desire to learn.

The defenders of rationality shake their heads at the pitiful gambler, whose perseveration is, we're told, sheer error. Gamblers suffer that most grievous of human maladies—cognitive bias, which causes them to miscalculate what they stand to gain from each irrational spin of the roulette wheel. Everyone knows that, on average, gamblers lose. To the rational observer, a casino's extravagant decor illustrates the house's ineluctable edge. To luck's naive acolyte, however, its luxury foretells the good fortune waiting just beyond one happy roll.

Yet this interpretation—reflecting economists' model of that relentless value maximizer, *Homo economicus*—doesn't quite add up. Casinos are a relatively recent invention, as are their massive payouts. For millennia, gambling happened on street corners, between friends. Without a middleman to bias payoffs, gamblers faced even odds. Gambling may have shifted money between players, or shuffled goods around, as it did in the pre-Columbian Americas. But on average, players broke even. Besides, if gambling addiction is driven purely by a gambler's expected payouts, why don't we ever find addicts gathered around vending machines, faithfully feeding them coins? Unlike slot machines, vending machines deliver rewards with a near-perfect success rate. Why is one so much more of a problem than the other?

Gambling's allure owes less to a player's misguided expectation of a massive payout and more to the pleasure of exploring the unknown. Early psychologists discovered that chance is, in itself, rewarding. Classic behavioral studies conducted by B. F. Skinner in the 1930s indicated that animals gamble much like humans. In Skinner's setup, rats learned to press a lever for a food reward. He rewarded some rats every time they pressed the lever. Others he rewarded only sometimes, like a slot machine payout. Skinner found that randomly rewarded rats came to compulsively press the lever. When Skinner stopped giving rewards for lever presses entirely, the rats who had received the random rewards continued to push the lever for much longer than the consistently rewarded animals. This is true throughout the animal kingdom—animals respond more strongly to uncertain rewards than to certain ones, even if the average expected gain is less. As neuroscientist Robert Sapolsky puts it: "Maybe is addictive like nothing else out there."

Neuroscientist Wolfram Schultz and his team discovered a similar phenomenon in their recordings from dopamine neurons. They trained primates to press a lever after a light cue to collect a juice reward. In their original experiment, dopamine neurons in trained animals were briefly active when the light flashed, signaling that the animals had learned to associate the light cue with a reward. In a modified experiment, the researchers trained the animals to expect a reward only some of the time. In this case, the dopamine neurons slowly ramped up their firing after the light cue was flashed. This rise was most dramatic when the reward was given 50 percent of the time. The dopamine ramp-up was smaller when the reward delivery probability was 25 percent or 75 percent. Dopamine release is highest when the

outcome is least predictable. Because dopamine release reinforces behavior that comes before it, this makes animals vulnerable to gambling
addiction. Studies in human subjects indicate that, in gamblers and
non-gamblers alike, dopamine release tracks a stimuli's unpredictability more closely than it tracks reward. The more unpredictable the
outcome, the stronger the motivation to gamble.

Psychologists Patrick Anselme and Mike J. F. Robinson suggest that
unpredictable phenomena elicit a strong dopamine response because
this motivates animals to persevere in the face of failure. Food foraging
is a skill necessary for all motile animals and is thought to have played
a fundamental role in the evolution of the brain. It is an inherently uncertain enterprise. As a result, the brain treats unpredictability with
particular sensitivity. In the wild, rewards are often random and rare.
Animals who give up easily are unlikely to survive. This explains the
counterintuitive finding that problematic gamblers often gamble even
harder after a string of losses, a well-documented phenomenon known
as "loss-chasing." The dopamine response motivates an unlucky forager to keep trying.

This is related to what's known as the explore/exploit trade-off. A
forager should know when to keep exploiting a winning strategy and
when to give up and search for something new. Evolution endowed our
minds with a fascination for uncertainty to help us more thoroughly
map our environment. People were drawn to understand chance, and
ultimately invent probability theory, by their biology. This also makes
us vulnerable to it: people and animals with dopamine disorders are
suboptimal foragers. Schizophrenic patients with disordered dopamine, for instance, are biased to explore more than exploit. A side effect of dopamine-enhancing drugs, such as those used in Parkinson's
patients, is gambling addiction—a tendency to overexplore.

Surprise can be a learning trigger, orienting an animal's attention
and sharpening its memory. It signals that there is something more to

understand. This may have been useful throughout evolution, incentivizing animals to explore novel stimuli and learn as-yet-unlearned associations. New information is valuable. Psychologists have found that people's brains respond to new information in the same way they respond to standard rewards like food and money. Knowledge is its own reward. Indeed, many pathological gamblers claim that they're driven by a feeling of being close to uncovering or understanding the patterns governing their games of choice. They believe that they have perfected, or are near to perfecting, a "system," or that they have internalized some deep insight into stochasticity. This illusion of almost-understanding can be intensely pleasurable. Gamblers report euphoric feelings in the trance of play, and winners often pour their earnings back into playing for longer. The goal is not to win but to extend the pleasure of playing as much as possible.

Uncertainty is a significant part of what makes any game enjoyable. Game designer Greg Costikyan notes that most adults don't play tic-tac-toe because they already know its winning strategies. Young children, on the other hand, still enjoy it because they haven't fully explored its outcomes. Uncertainty undergirds the pleasures of storytelling, of sports, of courtship. But it's also one of our deepest horrors: being blindsided by terrorist attacks, freak accidents, betrayals. Games sanitize and soften the realm of the unexpected, allowing us to explore and enjoy uncertainty in a nonthreatening way.

Games of chance were always latent in our neurobiology, waiting for the first person to link an external reward to the internal motivation to understand. When humans invented the earliest games of chance—likely involving the flipping of pebbles and knucklebones—they set the course for a deeper understanding of reality. Casinos are magisterial feats of social engineering, and they prey on humans where they are most manipulable: the realm of *almost*. With the right strategy, games like craps and roulette have a payoff probability hovering very close to 50 percent, optimally engaging the dopamine system. Designers have intentionally arranged the symbols on slot machines' reels to give players the sense of a near miss: I *nearly* had the complete line; I came *this close* to a jackpot.

Despite its vulgar reputation, gambling serves a beautiful impulse, reflecting the brain's compulsion to understand the world. The gambler's high is driven less by material greed than by epistemic greed. However, the brain is ignorant as to whether uncertainty comes from unpredictability in the environment itself (as with the unpredictability of dice) or from its own inability to work out the correct statistics of the world. We should be sensitive to this difference. Just as games can give players a sense of improvement and achievement in realms that have little bearing outside the game itself, gambling can provide an illusion of learning without genuine learning. Surprise is not a reliable proxy for genuine insight, as we know from the rise of clickbait and outrageous headlines. Ideas can be more valued for being surprising than for being right. Because surprising information engages our attention and memory systems, it is easier to remember and becomes naturally overrepresented in conversation. This is the seduction of the counterintuitive statistic. This is why contrarian figures like Jordan Peterson and Donald Trump attract so much attention. An economic essay may offer startling statistics that are eminently quotable at parties but bear little resemblance to reality. Surprising scientific findings are rewarded with publication in more prestigious journals, incentivizing scientists to chase (or manufacture) unexpected results instead of fleshing out established findings.

In many intellectual circles, counterintuitive facts or contrarian opinions are a status marker: interesting not because they're right or wrong but because they "make you think." Surprise is a heuristic for learning—but this doesn't always mean we're really learning from it. Rationality-focused economists and philosophers, often more interested in sophistry than genuine scholarship, delight in making the case that child labor is good, actually. Longtermist philosophers speculate on the outlandishly distant future with galling certainty in their ability to accurately estimate the probability of vanishingly unlikely outcomes, wielding probability theory as though it is an unassailable talisman against being wrong. Viewing the world through the filter of cost-benefit analyses, thinkers invent arbitrary values to assign as costs or benefits to prove their point. Longtermist philosopher William Mac-Askill argues, for example, that the collapse of biodiversity may not, in fact, be a moral loss:

> If we assess the lives of wild animals as being worse than nothing on average, which I think is plausible (though uncertain), then we arrive at the dizzying conclusion that from the perspective of the wild animals themselves, the enormous growth and expansion of *Homo sapiens* has been a good thing.

He might as well argue that we can assume that any underprivileged person who speaks a language we don't understand is better off dead. Therefore, as this logic goes, there'd be no harm in killing poor immigrants. This is only one recent example in a long line of thinkers employing transparently delusional probabilistic reasoning to make specious points. Through seemingly objective arguments like these, the status quo keeps its finger tipping the scales in the marketplace of ideas, assigning invented values to justify arbitrary ideologies.

Probability theory was a wholly new form of math, so unlike that which came before it, twentieth-century scholars undertook a concerted effort to tie it back to the rest of the field. It transformed mathematics, corroding the empire of logical inevitability with the force of *maybe*. It has since come to guide and clarify human decision-making in finance, industry, and the sciences. In place of the old reliance on intuition and tradition, probability theory allows people to quantify risk and uncertainty and more accurately predict—even control—the outcome of their decisions. It allows us to imperfectly excavate the future, laying out possible outcomes and estimating which are most likely. Intelligence has been defined as the ability to navigate these potential futures to arrive at the optimal one—an idea at the heart of gameplay. Some have argued that thought itself works according to statistical logic.

Chance is everywhere we look and fundamental to the nature of matter itself. It is the ingenious language we use to express our ignorance, and one that served as a powerful foundation for the rise of empiricism. Unfortunately, its illusory rigor can be misleading and can cause us to believe that we know more than we do by assigning numerical values to our ignorance. But the study of games of chance indicated that by investigating something widely enjoyable, researchers could better understand how the mind works, how people work. This insight would bear fruit with the development of game theory and the game-inspired development of artificially intelligent systems.

It is hard to overstate the magnitude of this discovery and how profoundly it has changed the way humans reason about the world around them. Games revealed that some forms of chance—once emblematic of the unknowability of God's mind—are governed by laws. These laws had, for millennia, remained hidden by the very noise they emanate. By contemplating dice, people wrested control of their future from fate and turned the former art of decision-making into a quantifiable science. Or so they hoped.

Part II

Naming the Game

4
Kriegsspiel, the Science of War

Play is the beginning of knowledge.

GEORGE DORSEY

t was a military technology so quietly powerful that even its creators failed to grasp its full potential. Built over centuries of collaboration by prominent scholars and scientists, it was perfected by a father and his son: Georg Leopold von Reisswitz and Georg Heinrich Rudolf Johann von Reisswitz. In 1824, the younger Reisswitz, a Prussian military officer, blithely demonstrated their invention at the residence of the future tsar of Russia, then a German ally. Over the course of the next hundred years, this invention would transform the very nature of modern warfare and alter the face of global politics. In the nineteenth century, it helped anneal the first German Empire out of long-warring states. In the twentieth century, its failures fractured the empire and thwarted the leaders' global ambitions.

A botched medical procedure prevented the elder Reisswitz from becoming an army officer, as had long been his family's tradition. Instead, he served Prussia as a military strategist, spending decades developing a device that was not a weapon but a means of cultivating thought—a strategy for inventing strategies. It's a method still used today by every modern military organization: Kriegsspiel, or "war game." The Reisswitzes demonstrated that games could do more than aid our thinking in the abstract; games could help us simulate and predict reality.

Kriegsspiel had been inspired, in part, by the seventeenth-century mathematician Gottfried Wilhelm Leibniz, who'd self-funded a short-lived Berlin academy focused on games and technology. Leibniz was born around the same time as probability theory, which was, much like AI today, the new science that scholars promised would solve all modern problems. In part because probability theory had its provenance in games of chance, games were increasingly regarded as subjects worthy of serious academic interest. Leibniz saw a deeper significance. The universe, he claims, is rather like a game of chess. God had conceived of all possible worlds in a way akin to the space of all possible chess moves. Of these, only some subsets are legal, playable moves—what Leibniz calls "compossible." And of these many potential games, only one, deemed by God to be the best possible, came into being as the world we live in today. Below the threshold of expressed reality, a thousand potential worlds bicker and battle for material existence. It is a strange kind of thought, Leibniz admits, but games are "the most close representation of human life."

Games, in Leibnizian philosophy, are the language through which God expresses the universe. As such, humans can use games to understand how reality is composed and how to craft better futures. Leibniz believed that games could be harnessed to generate new insights into areas ranging from military tactics to medical treatments. More than that, games are a form of meta-learning. "I strongly approve the study of games of reason," he writes, "not for their own sake, but because they help to perfect the art of thinking." In the years since Leibniz's time, games and simulations have become standard thinking aids in humanity's groping toward perfection. Unfortunately, they've been as helpful in determining the best possible war as they have been in navigating toward the best possible world.

Kriegsspiel roots back to the late eighteenth century, when mathematician Johann Hellwig invented a variant of chess. Chess was then considered to be required training for nobility in Europe and the Near East; an aristocrat was expected to master calligraphy, horseback riding, history, philosophy, weaponry, music, astrology, and games, especially chess. Tutors used chess to teach strategic thinking and mathematics to their pupils. Chess, originally chaturanga, originated in India as an explicit abstraction of war. Its name, which means "four limbed," refers to the four ancient army divisions of infantry, cavalry, elephantry, and chariotry. In the Arabic world, the pieces evolved from lifelike sculptures of men, horses, and elephants to the more abstract forms we know today, filtered through the Islamic prohibition on figurative art. As the game percolated along trade routes, elephants transformed into camels in the East and bishops in the West. In Europe, where powerful queens were on the rise, the vizier piece came to be replaced by the more dynamic queen.

Hellwig worked to add military realism back into chess, an ambition that would turn the game into a more expressive war computer. He replaced the pieces representing outmoded military units with modern artillery and substituted the king with a fixed fortification: armies are often more interested in capturing cities than kings. To mimic a real battlefield, Hellwig expanded the board to 1,617 squares, each color coded to indicate different terrains. Otherwise, the game's mechanics remained similar to those of chess. Hellwig and other game enthusiasts continued to develop new variations on war chess, adding features that would become standard in war simulations. They used dice to determine an attack's damage, reintroducing the noisiness of reality into chess's composed strategy. Still, military officials overlooked Hellwig's game. Their battalions, after all, had never marched through a flat-grid world, or fought to the last man, or met many of the other simplifying assumptions of the game.

In the early nineteenth century, the elder Reisswitz foresaw the game's potential for military training and began iterating on increasingly elaborate versions. He replaced the chessboard with wet sand molded into imaginary landscapes but hated to think of the royal family playing with such inelegant stuff. Next, he designed porcelain tiles with bas-relief terrain features. Players could arrange these in modular configurations and virtually experience the difficulties of different terrains. Wooden blocks represented troop formations. Reisswitz gifted his Kriegsspiel to the king and organized a grand demonstration. It became a favorite plaything of the young princes but was otherwise neglected.

As he advanced through the military academy, the younger Reisswitz grew convinced of the promise of his father's vision. By the time he took over Kriegsspiel's development, advances in mapmaking techniques meant that it could be played on scaled maps of actual battlefields. Recent advances in statistics enabled the younger Reisswitz to introduce a data-driven scoring system, inspired by the empiricism that was then the rage in Europe. He made the game more rigorous, employing odds-based score tables using data from historical battle losses. (Later, Don Rawitsch borrowed this technique in his design of the early video game *The Oregon Trail*, mining the diaries of early American pioneers to set the probabilities of bandit attacks, cholera epidemics, and river crossings.) Dedicated dice for cavalry, infantry, and artillery determined the damage of an attack. The two sides were designated by red and blue, a convention that persists to the present day—security professionals use "red teaming" to simulate attacks on their systems. Each turn simulated two minutes of warfare. The wooden-block troops, which took up map space corresponding to the actual footprint of regimental units, were constrained to move a reasonable distance therein. Communication between the forces was subject to the fog of war, and opposing players used separate boards so that they couldn't see one another's pieces, just as real intelligence is limited by troops' relative positions and sight lines. Kriegsspiel systematized contingencies of terrain, weather, and enemy power.

In 1824, Prince Wilhelm—who had loved the game's earlier incarnation as a child—had learned of the improved version through his military tutor. He invited the younger Reisswitz to umpire a demonstration of Kriegsspiel at his home, witnessed by princes, dignitaries, and the king. The game took weeks to play. During that time, all cats had to be banished from the house to prevent them from disrupting the game pieces. The scenario involved a campaign fought between the Oder and Elbe Rivers, including territory in what is now Poland—a region of great strategic interest at the time. General Karl Freiherr von Müffling, who attended the demonstration, quickly realized the revised game's potential and exclaimed: "It's not a game at all! It's training for war. I shall recommend it enthusiastically to the whole army." The king was likewise impressed and demanded, by royal decree, that every military regiment learn to play.

Despite the game's apparent power, it was hardly kept a military secret: books on Kriegsspiel grew popular across Europe and were translated into several languages. The younger Reisswitz visited St. Petersburg for a summer to teach the game to Grand Duke Nikolaus, a kindness that would doom German forces one hundred years later. In the meantime, the young officer suffered for his success. Though the game was a favorite pastime of the royals and had support from some high-ranking military officials, it was unpopular among his fellow officers. Kriegsspiel was complicated and its rigid rules challenging to learn. Many officers resented the days or weeks they lost to its tedious simulations. Reisswitz's service as an umpire was particularly galling to higher-ranking officials, who hated being subjected to their subordinate's judgment. In 1826, the younger Reisswitz was assigned to a backwater military post by colleagues jealous of his success and proximity to the royal family. Reisswitz interpreted this assignment as exile and shot himself not long thereafter. His heartbroken father died a year later.

The same Prince Wilhelm who had loved Kriegsspiel as a child went on to serve in the military. He became a celebrated general, thanks in part to the game's honing of his strategic genius. The death of his childless elder brother in 1861 made Wilhelm king, but the brilliant military tactics devised in Kriegsspiel's simulation engine made him emperor. For the first time in its thousand-year history, Germany had a cohesive national identity, which Wilhelm had shaped out of a disorganized "nerveless body" of local states. His military campaigns stunned the world with their power. Dulled by decades of his brother's peaceful reign, his generals were re-sharpened on Kriegsspiel's whetstone. Prussia's unexpected and swift win in the Franco-Prussian War was widely attributed to their gamed simulations. Other countries took note.

General Helmuth von Moltke, who would go on to be the chief of staff of the Prussian Army, was a devoted fan of Kriegsspiel. He saw it as an ideal training tool for officers. Anyone, he argues, "can take part immediately in the game as a commander, even if he has no previous knowledge of the game or has never even seen it before." He used the game to invent some of his more famous maneuvers and credited the game's bird's-eye view of the battlefield for offering a novel perspective on outdated tactics. For instance, Moltke abandoned column and line troop formations, which he realized made for easy targets, in favor of loose formations. The lower density of soldiers translated to fewer casualties in artillery attacks.

Earlier versions of the game had been tedious to play, each turn burdened with complicated formulas for tallying the exact numbers of casualties. In time, the designers replaced the onerous rule book with knowledgeable umpires who could quickly estimate and assign damage or assess victory after each move. Kriegsspiel became significantly faster paced and even—for some—enjoyable. The game made it possible for military officers to improve their skills even in times of peace. It combined the pure strategy of chess with the chance element of dice, so players had to be both future oriented and quick to adapt. Moltke organized the first permanent peacetime general staff and had them play Kriegsspiel to keep sharp. It allowed officers to assess their reports' intelligence and strategic savvy and generally disrupted the military's hierarchy. Previously, appointments had been nepotistic, but Kriegsspiel made it easier for officers to ascend through the ranks on the merit of their ideas. Trainees could reason through scenarios with their superiors, coordinating on high-level strategy and working out counterfactual outcomes.

Officers were afforded more accountability, and game simulations gave them a deeper understanding of the tactics they should carry out. Prussian general Kraft Karl August zu Hohenlohe-Ingelfingen, whose military strategy won him high honors in the Franco-Prussian War, credited his success to Kriegsspiel: "The ability to quickly arrive at decisions and the cheerful assumption of responsibility which characterized our officers in the Franco-Prussian War of 1870–71 was in no small measure due to the war games." Details of a tactical plan could be workshopped, rehearsed, and procedurally embodied in the decision-making of combatants at all levels. Decision-making was decentralized and relegated to lower-ranking officials, who had direct information about the conditions at hand. Through practice, officers harmonized their efforts and became versed in one another's strengths and weaknesses. Well-trained and aligned military personnel could make decisions as small units and move quickly. The overall strategic vision was communicated programmatically rather than as a static order: *If* x, *then* y. Through the lens of a game, war became a science: Testable. Rigorous. Algorithmic.

Before its revival under Kaiser Wilhelm, Kriegsspiel had been largely forgotten in the long peace that characterized the reign of his elder brother. Scholars had dropped the disgraced Reisswitz name from any mention in military histories and Kriegsspiel manuals. In fact, the Reisswitzes had been so effectively scrubbed from history that an 1873 magazine article celebrating Kriegsspiel's role in Prussia's recent military victories failed to mention the family whatsoever. The author instead claimed that the game had no inventor and had been passed down as folk knowledge until the first guide was published in 1846. A few weeks later, an anonymous letter to the magazine corrected the record, describing how the game had been painstakingly perfected over decades of experimentation by the Reisswitzes, and how it had come to the attention of the royal household. Though the letter was unsigned, most historians believe that it could only have been authored by the emperor himself, given that it contained details known solely to him. It was an act of patient tribute to the father and son whose technology had forged a national identity for Germany and changed the face of Europe—and, soon, the world.

The German military continued to update Kriegsspiel with the latest data on casualties from ever-improving weaponry, and the game continued to serve as an uncannily accurate prediction engine for Germany's aggressively expanding army. It allowed strategists to explore an array of hypothetical scenarios—like Leibniz's chess-playing God, like the brain's decision system—and investigate the space of possibility before choosing the best option. Officers used simulations not only to test specific tactics but also to lay out entire war plans. Helmuth Johann Ludwig von Moltke, nephew of the elder Moltke, became the army's chief of staff shortly before Germany entered the fray of World War I. Germany was now pitted against Russia, its former ally. Like his uncle, Moltke relied heavily on Kriegsspiel to plan everything from high-level strategy to mundane operations. When used to game out a critical battle, Kriegsspiel indicated that troops on the far-right flank would run out of ammunition two days before the campaign ended. Moltke therefore organized motorized ammunition battalions to replenish the troops; these were the first of their kind and crucial to the campaign's success. Kriegsspiel was the first supply-chain forecasting program.

However, World War I would highlight what was missing in these simulations. Moltke and his predecessor, Alfred von Schlieffen, used Kriegsspiel to generate a holistic war plan. Despite carefully gaming out an invasion of Belgium, they misjudged its social and political repercussions. Moltke and von Schlieffen were taken by surprise when Belgian civilians tore up their own train tracks to make it harder for the German invaders to transport supplies. They had also failed to consider the diplomatic fallout of the invasion. Belgium had been neutral since 1839. Germany's attack drew Belgium's ally Britain into the fray and, later, the US, ultimately leading to Germany's defeat.

The Treaty of Versailles, which brought an end to World War I, imposed significant penalties on the fallen German state, including severe restrictions on its military forces. Its standing army could not exceed one hundred thousand troops, nor its navy more than six battleships. But Kriegsspiel enabled German officers to continue to train the skeletal troops, strengthening their skills virtually. Officers prohibited from interacting with actual troops could test the potential outcomes of hypothetical decisions on the game board. German military scholars also used Kriegsspiel to reflect on their losses, dissecting historical battles to understand why they'd succeeded or failed. The historian Hans Delbrück recognized that military and political history are inextricably tangled. He was committed to establishing a quantitative study of history. Throughout his career, Delbrück pored over historical data, hoping to develop a science of war, relentlessly work-shopping historical battles to test his theories. He argued that Germany's failures in World War I were primarily strategic in nature. He advocated for their neutered military to take up historical study and simulations. By embracing these simulations during the interwar period, the Germans developed a new form of combat using theoretical forces that did not yet exist but soon would. It promised to solve the stagnant trench warfare that had characterized World War I, leveraging surprise to dominate and demoralize the enemy. In World War II, this tactic would rain terror from the skies across Europe. It would come to be known throughout the world as the Blitzkrieg.

Adolf Hitler was made chancellor of Germany in 1933, just fifteen years after the end of World War I. Though he promised German citizens a new empire by military conquest, he was generally unimpressed with Kriegsspiel, preferring his gut instincts to its cold calculations. His officers nevertheless continued to mock up scenarios behind the scenes, unwilling to abandon the technology. Lieutenant Colonel Erich von Manstein used Kriegsspiel to demonstrate possible outcomes of an invasion of Czechoslovakia, and he convinced Hitler to commit to a strong initial offensive force, contrary to the leader's original instinct. These and other simulations enabled the Germans to dominate their neighbors faster than might otherwise have been possible. Next, Kriegsspiel correctly predicted that two battles against Britain would end in draws. Hitler began to take the game more seriously.

The dictator had set his sights on the Soviet Union, hoping to expand his empire and secure land for his "master race" to colonize. His generals staged a Kriegsspiel simulation dubbed OTTO. They spent three weeks workshopping the invasion. In retrospect, the German military's projections for its Russian campaign were shockingly accurate. The game indicated that German troops would destroy 240 Soviet divisions by November, leaving only 60 divisions remaining. The generals didn't bother extending the simulation further; their predicted margin of victory was so large that the Red Army would be effectively extinguished. The campaign went ahead largely as expected. On the day Germany's corresponding simulation ended, its troops had advanced exactly as far into Russia as the game had prophesied and had destroyed 248 Soviet divisions—eight more than anticipated.

But the Soviets had also war-gamed the Axis offensive. The Russians had adopted Kriegsspiel after the younger Reisswitz presented it to Grand Duke Nikolaus a century earlier. In the Soviet simulation of the invasion, General Georgy Zhukov, a rising military star, had role-played brilliantly as the Germans. Soviet leader Joseph Stalin was so frightened by the success of Zhukov's simulated campaign that he ordered his generals to throw everything they had at defending against the German offensive. Though the Russians had lost 248 divisions to the invading Axis powers, Stalin had mobilized so many troops that 220 divisions, not 60, remained to fight off the Germans. And winter was approaching—a factor the Germans had failed to consider in their abbreviated simulation. They'd declared imaginary victory too soon. If the Germans had continued to workshop OTTO past November, they might have been better prepared for the devastating winter that was to come. Though the war would grind on for several more years, this disastrous defeat marked the beginning of the end for Hitler's Reich.

Kriegsspiel soon reached beyond the narrow confines of officers' clubs and history departments. Prominent thinkers fretted about the rise of democracy across Europe, worried that the average person, agitated by nationalism, might vote in favor of unnecessary wars. War games were offered as a preventative—they might become a crucial tool for educating the populace on the difficult trade-offs of war. The science-fiction writer H. G. Wells published a book of rules in 1913 detailing a Kriegsspiel-inspired tabletop war game he'd invented. It was condescingly named *Little Wars: A Game for Boys from Twelve*

Years of Age to One Hundred and Fifty and for That More Intelligent Sort of Girl Who Likes Boys' Games and Books. Wells was an avowed pacifist, and throughout the book he took pains to calm his conscience, avowing that the game would experientially demonstrate the true horror of conflict. Britain was transforming into a democracy, and Wells hoped to appeal to the masses to ensure that they would never vote in favor of war. "If we do not end war, war will end us," he despaired. *Little Wars* was intended as a prophylactic, meant to inoculate its participants against violence. It would not: Winston Churchill, for instance, was an avid player of Wells's war game.

By the mid-twentieth century, war-based games had found popularity in the US—Europeans, understandably, were less interested in playing military games, having faced devastation on the ground. Americans, shrouded from the war arena, played at toy battles sanitized of death camps, blitzkriegs, and suicide attacks. Kriegsspiel spawned a new family of war-based tabletop and board games, including Warhammer, Chainmail, and countless others—even games like Settlers of Catan borrow some elements of war play. In the 1974 hit Dungeons & Dragons, individual characters replaced platoons, and a dungeon master replaced the umpire. As with Kriegsspiel, dice rolls were used to determine the damage related to attacks and gambits. These games served as the foundation for text-based, then graphical, video games in which programming took over the role of the umpire and computer-generated random numbers replaced dice rolls.

While Kriegsspiel changed the face of modern combat, war and games share a long history. Anthropologist Michelle Scalise Sugiyama believes that team sports emerged as a way for people to practice the skills necessary for war. Her research group identified movements common in sports—catching, dodging, grappling, kicking, parrying, running, striking, throwing—and found that these are also common in raiding activities. What's more, in many cultures, games were used in place of war. For millennia in Mesoamerica, the Olmecs, Mayans, and Aztecs used a ballgame to settle disputes. The game, usually played by members of the aristocracy, was used to consolidate power and clarify territorial boundaries. It was so important that a central Mayan creation myth in the Popol Vuh, the foundational sacred book of the K'iche' people, revolves around a match played between mortal heroes and the gods of the underworld. Other American games, like chunkey and lacrosse, were similarly used as peaceful proxies for war. These games were often preceded by the same rituals that preceded war. Some scholars have argued that lacrosse was crucial to maintaining peace and coherence among the Six Nations of the Iroquois. In the present day, sports remain a cohesive force, uniting groups locally and nationally.

Kriegsspiel served as a material scaffolding for thought. It empowered players to reason collectively to generate novel tactics and testable predictions. It was a world in miniature, its players minor gods whose dice rolls decided the fates of thousands of people, both real and imagined. Chess was transformed into an astonishingly accurate simulation engine by modifying its rules and pieces. To this day, all major militaries still use some form of Kriegsspiel-style simulation to inform their decisions. The game helped military theorists make explicit the conditions of battle and gain new perspectives on war. Soon, a further abstraction of games would help thinkers model *any* form of interaction, whether adversarial or cooperative. Games would move beyond their role as thinking aids and inspire a mathematical language in which the nature of conflict itself could be expressed.

Of course, reality can't be easily simulated as a game. Take Moltke and Schlieffen's failure to account for diplomatic repercussions when they gamed Germany's invasion of Belgium in World War I. The attack was feasible on paper but ultimately provoked the neutral coun-try's allies to enter the fray. It is the nature of models to be simplifica-tions of the systems they emulate. As the German military had discovered in both world wars, what's left out of a game can be just as important as what's included.

5

Rational Fools

But it is clear that we must embrace struggle. Every living thing conforms to it. Everything in nature grows and struggles in its own way, establishing its own identity, insisting on it at all cost, against all resistance.

RAINER MARIA RILKE

C hess is the bestselling game of all time and currently boasts an estimated six hundred million players worldwide. Over the fifteen hundred years of its existence, it has been so well loved that it has been routinely outlawed by religious leaders, kings, and caliphs threatened by its popularity. By the nineteenth century, the top chess players in Europe were cultural icons, and like modern-day influencers, they boosted their personal brands by selling strategy books and celebrity appearances. Chess clubs and magazines were commonplace. Chess was a lingua franca spoken by kings and mendicants, soldiers and civilians, artists and scientists alike.

From the eighteenth until the late nineteenth century, Romantic-style chess was the prevailing fashion, characterized by pomp and drama. Players brashly sacrificed pieces and won in daring last-minute maneuvers. Flair trumped success. By the end of the nineteenth cen-

tury, however, world champion Wilhelm Steinitz dominated his opponents with a more strategic style. He plotted positional gambits over
long time horizons, focusing on controlling the center of the board to
leverage possibility. A centrally located piece simply has more potential moves to make. This style was perfected by Steinitz's successor,
Emanuel Lasker, one of the most dominant chess players in history.

Lasker, the son of a Jewish cantor, was born in present-day Poland.
When he was eleven, his parents sent him to Berlin to study mathematics under the care of his older brother, Berthold. Berthold was one
of the top ten chess players of the 1890s, but Emanuel soon outshone
him. By age twenty, he'd emerged as the world champion. He abandoned mathematics for this more lucrative profession and remained
the world chess champion for twenty-seven years, from 1894 to
1921—the longest reign in history. Like Steinitz's style, his unorthodox mode changed the face of gameplay. The fact that chess was subject to changes in fashion surfaced important questions: If some play
strategies were better than others, was there such a thing as a *best*
strategy? Was chess, once valued for its unpredictability and seen as
an infinite horizon of spontaneous yet unique encounters, in fact
knowable—calculable?

Albert Einstein admired Lasker and believed he was after something bigger than chess. "What he really yearned for," Einstein surmised, "was some scientific understanding and that beauty peculiar to
the process of logical creation, a beauty from whose magic spell no one
can escape who has ever felt even its slightest influence." Lasker wrote
volumes on the theory of chess play but also some works of philosophy,
most notably his 1907 book: *Kampf* (*Struggle*). He intended *Kampf* to
be a general theory of competition, encompassing games, business, and
war. In it, he draws connections between gameplay and society, linking
the struggle of players with the choices of *Homo economicus*—the rational agent of economists' models. "What is struggle and victory?" he

writes. "Do they obey laws that reason can comprehend and formulate? What are these laws? That is the problem!" Games were a perfect test bed for this endeavor. He thought that scholars might one day express theories of chess play mathematically. "Humanity stands before the gate of an immense new science which prophetic philosophers have called the mathematics or the physics of contest," he writes.

This notion of struggle, he contends, is applicable to not only people but also nations, the natural world, and even languages. He believed that struggle would come to be understood as a measurable quantity: *I endured fifty nanowars arguing with my spouse today*, perhaps, or *These new regulations will cost four centibattles of work for the legal team*. The study of games and conflict would eventually yield a new physics—for example, the "struggle" performed by bullets in a war might be measurable by the number of bodies they felled. Even atoms, he argues, are subject to struggle, helpless to the attractive and repulsive forces between their constituent protons and electrons. This meant that conflict could be predictable: entities struggling against one another should fall into a stable equilibrium, like an electron constrained to its orbital cloud. Lasker felt that the study of struggle would transform politics—and perhaps even end war—because its theorists would discover alternative ways to settle disputes. "That wars should appear to us a necessity is proof of our stupidity," he writes. Games would help scientists frame the mysterious contests of life in precise terms and determine their solutions. People once solicited the aid of gods before making critical decisions, he surmises, because they believed that "neither reason nor justice but the dictates of an autocratic power governs destinies." He hoped this new science could usurp the tyranny of fate and tame the future with the yoke of reason. A new mathematics of games was just on the horizon. It has not, unfortunately, come to end war.

Games continued to draw scholarly interest. Psychologists wrote

Freudian interpretations of chess, reading Oedipal passions into the ambition of rendering the king, a father figure, helpless and impotent. "The unconscious motive actuating the players is not the mere love of pugnacity characteristic of all competitive games, but the grimmer one of father-murder," writes Freud's disciple Ernest Jones. Mathematicians, conversely, sought a purely analytic treatment of games, stripping away all the psychology of game players. The German mathematician Ernst Zermelo presented a paper in 1912 analyzing whether one could objectively determine how good an arbitrary chess arrangement was. What constitutes a winning position, and can that be mathematically defined without resorting to subjective or psychological concepts? An unfinished game of balla, as Pascal and Fermat elaborated, has only a handful of possible outcomes. Similarly, an incomplete game of chess might resolve in a finite—though huge—number of ways. Zermelo introduced a new way to think about games: as collections, or sets, of possible moves that eventually lead to wins or draws. Chess, considered by some to be the zenith of human intellectual achievement, could be reduced to a search problem—less an art than a massive decision tree.

A decade later, the French mathematician Émile Borel took poker, instead of chess, as his muse. Unlike Zermelo, Borel was interested in the intersection of psychology and play. Chess is a game of perfect information: both players know where all the pieces are at any given time. In contrast, poker players take great care to hide the strength of their cards, sometimes through deceit. Bluffing often decides the game. Borel proposed a program he calls a "method of play," similar to what would come to be known as a "strategy." This, he writes, is "a code that determines for every possible circumstance [. . .] exactly what the person should do." Throughout the 1920s, Borel worked out optimal strategies for games including Rochambeau. A "pure" strategy would look like a person always playing rock, always playing paper, or always playing scissors. But this would be easily exploited by their opponent, who would, accordingly, always play paper, always play scissors, or always play rock. For Rochambeau, a "mixed" strategy is optimal: opponents should play rock, paper, and scissors randomly, with equal probability. A player using this strategy won't always win, but they will minimize their losses. Borel believed that the psychological and subjective aspects of gameplay would necessarily limit what mathematicians could say about games, so he eventually abandoned the research.

The Hungarian mathematician John von Neumann would eventually strip psychology from games entirely, rendering them pristine mathematical reflections of the surface of all possible interactions. Born in 1903 in Budapest, von Neumann was immediately recognized as a prodigy. He came from a wealthy and recently ennobled Jewish family. They'd appended the honorific *von* to their surname when John was ten, as the cash-strapped leader of the Austrian Empire, Franz Joseph, sold nobility titles to the rising mercantile class to fundraise for the military. Von Neumann had a photographic memory and was known to recite pages from the phone book as a party trick to impress his parents' friends. He could multiply eight-digit numbers in his head and tell jokes in ancient Greek. He loved playing games, particularly Kriegsspiel, which matured into a general interest in war itself. When World War I broke out, the young von Neumann kept a record of troop movements, sketching the shifting battlefronts on carefully annotated maps.

Von Neumann's intellect was legendary even among the legendary. The mathematician Eugene Wigner claimed: "Whenever I talked with the sharpest intellect whom I have known—with von Neumann—I always had the impression that only he was fully awake, that I was halfway in a dream." The great mathematician George Pólya recounted a time he had taught von Neumann at university:

Johnny was the only student I was ever afraid of. If in the course of
a lecture I stated an unsolved problem, the chances were he'd come
to me as soon as the lecture was over, with the complete solution in
a few scribbles on a slip of paper.

Throughout the 1920s, von Neumann published groundbreaking
work in fundamental mathematics and the foundation of quantum
physics, where he grappled with the probabilistic nature of atoms. To
this day, that remains the contentious knot at the core of conflicting
interpretations of quantum theory. In 1926, by the age of twenty-
three, von Neumann had earned a degree in chemical engineering
from the Swiss Federal Institute of Technology in Zurich and his doc-
torate in mathematics from the University of Budapest. That same
year, he presented, almost as though an afterthought, a lecture on
games that contained the germ for the field of game theory.

The common thread binding von Neumann's diverse interests can
be attributed to the influence of his graduate adviser, the mathematician David Hilbert. At the turn of the twentieth century, Hilbert had
challenged the world with twenty-three grand problems. His sixth
challenge was to axiomatize all of physics and mathematics, or reduce
them to a set of axioms from which a consistent, self-contained system
of knowledge could be derived. Axioms are self-evident propositions.
The Greek mathematician Euclid had derived the entire field of Eu-
clidean geometry from five axioms, or simple observations—for instance, that a line can be drawn bridging one point to any other point.
Hilbert wanted mathematics to be torn down and rewritten from the
ground up, in a manner so careful that it left no room for uncertainty
or contradiction. Like a well-designed game, mathematics should con-
sist of minimal rules from which a maximally expressive system could
be built.

Hilbert argued that math should be founded on pure numbers,

which are an internally consistent realm. Numbers don't refer to anything outside themselves yet obey clear rules governing their relations. Numbers, he claimed, come before logic, existing "intuitively as immediate experience before all thought." One plus one, inarguably, equals two. The number line can be assembled through the simple operation of adding new strokes to a string of tally marks. Mathematics, once a product of human intellect, would instead be constructed by mechanical procedures, each step proceeding inevitably according to the system's internal logic. It would reflect, as its premises unfolded to their inexorable conclusions, a purely mechanical rationality, sans thinking. Some accused Hilbert of denuding mathematics of any real meaning, making it instead a play of abstract symbols. His student Hermann Weyl writes: "Mathematics is no longer knowledge but a game of formulae, ruled by certain conventions, which is very well comparable to the game of chess." But Hilbert defended his program against accusations that mathematics would degenerate into a game. It wouldn't obviate thought. It was, instead, an expression of how thinking works. "The fundamental idea of my proof theory," Hilbert asserted, "is none other than to describe the activity of our understanding, to make a protocol of the rules according to which our thinking actually proceeds."

To Hilbert, the entirety of reality was simply waiting for our discovery. It issued from the same self-evident simplicity that makes one and one equal two, just as all possible chess games unfold as the mechanical consequences of its rules. He believed that truth could be fully intuited and needn't ever connect to anything in the world of experience. Hilbert's program inspired an "axiomania" that had academics attempting projects ranging from uncovering the logical structure of language to modeling psychology as a kind of latent inner geometry. He'd hoped to challenge academics to systematize not only pure math but also statistics, physics, and other sciences, forcing fresh blood into distant fields still untouched by the articulating force of mathematics. To this day, Hilbert's challenge continues to echo down through the code bases of academic institutions and major corporations, reflected in procrustean efforts to mathematize social network dynamics, algorithmically detect emotions, and predict complex emergent features like consumer choices, crime recidivism, and dating preferences by sieving hidden relationships from big datasets.

As part of these efforts, Hilbert's advisee von Neumann hoped to axiomatize human nature. Johnny was known, as his brother Nicholas recounted, for his ability to find mathematical solutions to problems that "a priori do not seem amenable to mathematical treatment." Nicholas believed that their mother had instilled this habit, as she'd always delighted in stories of people, like the Antarctic explorer Ernest Shackleton, attempting the impossible. Her father, after all, had lifted himself from rags to riches, a feat she was fond of relating. Johnny tried to perform another sort of miracle: transmuting humans into mathematical objects. He did this by modeling them as players in a game, their decisions actuated by their desires.

Few others could bring the expressive force and mathematical rigor that von Neumann brought to games. In 1926, he presented a paper describing players in any finite two-player zero-sum game. Zero-sum games are games of pure conflict—like chess or poker—wherein gain for one player necessarily means an equal loss for their opponent. There are other kinds of games: In a positive-sum game, play increases the total value shared by players—for example, collaborating on a project at work that scores all participants a promotion. A negative-sum game, like war, results in value being destroyed.

Von Neumann proved that, for all two-player zero-sum games, there exists a strategy that allows each player to minimize their opponent's gain. This means there is a predetermined, if pessimistic, best outcome that both players can hope to achieve, no matter how well their opponent plays. "It is easy to picture," von Neumann writes, "forces struggling with each other in such a two-person game. The value of [player one's winnings] is being tugged at from two sides, by [player one], who wants to maximize it, and by [player two], who wants to minimize it." Each player's goal should be to minimize their opponent's maximum payoff. In a zero-sum game, this is the same as maximizing their minimum win. This value, known as the minimax, represents the highest payoff a player can hope for—even in the worst-case scenario and knowing nothing about their opponent's approach. A rational player will want to use this strategy, no matter what their opponent does. Assuming both players are rational—that is, they seek to maximize their payoffs—both will gravitate toward the minimax strategy. It is an equilibrium of the system. The game's outcome is then entirely determined by the rules of the game, not by the psychology of the game's players. All the complexity and drama of play can be boiled down to a game's equilibrium points. Much as the gravitational pull of massive bodies enforces the orbits of planets, so the players' self-interest impels their use of the minimax strategy.

This is known as the minimax theorem. Minimax might strike some readers as overly conservative: Surely players can hope to do better than merely minimizing their maximum losses, right? Can't they aim to maximize their gains, for example? This is known as the maximax strategy. Maximax, it turns out, is overly optimistic and ultimately irrational, often adopted by naive players and young children. It's irrational because it doesn't account for the opponent's strategy. Take the classic parenting trick for dividing a dessert between two siblings: one is tasked with cutting the cake, and the other gets to choose their piece first. This setup incentivizes the cutter to split the cake as fairly as possible, because they know their sibling will take the larger slice. This is a minimax strategy. But a sibling employing a maximax strategy would cut a tiny sliver from the cake and hope against hope that their sibling chooses this piece, leaving them the large remainder.

Von Neumann, like Lasker, recognized a parallel between game dynamics and the social sciences, though his paper only hints at this connection. He also alludes to the notion of struggle, stressing that almost any interaction can be characterized as a game. In a footnote, he draws a tenuous connection between his work and economics, likening strategies to "the principal problem of classical economics: how is the absolutely selfish 'homo economicus' going to act under given external circumstances?" Minimax would later become a foundational concept not only in the social sciences but also in computer science, decision theory, AI, and ethics.

In the same way it once might have struck people as preposterous that chance could be quantified, the notion that mathematics could strip psychology from gameplay was alien. "It makes no difference which of the two players is the better psychologist, the game is so insensitive that the result is always the same," von Neumann writes. His work reflected Hilbert's dream: math would eventually imperialize all fields of knowledge. No inquiry would remain resistant to quantification. However, von Neumann would temporarily abandon this result, not returning to it for almost a decade.

By 1928, mathematicians had made enough progress on Hilbert's axiomatizing program that Hilbert declared it nearly complete. There were only a few loose ends to tie up. Von Neumann had contributed several important advances, and his colleague Stanislaw Ulam remarked that von Neumann "seem[ed] to realize Hilbert's goal of treating mathematics as a finite game." Two years later, however, the logician Kurt Gödel presented a revolutionary proof at a conference in Königsberg. His manner of presentation was so understated that his results seemed to go largely unnoticed. Von Neumann pulled Gödel aside afterward to ask whether he had understood rightly. Gödel had discovered that formal systems can't demonstrate their own consistency. There is no way to define a set of axioms for all mathematics. Mathematics will always be incomplete because sometimes true statements cannot be proven. This is often illustrated with the liar's paradox: "This statement is false." If this statement is true, then it's also false—and is therefore neither. Gödel showed that the same can be true of propositions in mathematical systems. "My personal opinion," von Neumann later reflected, "is that Gödel has shown that Hilbert's program is essentially hopeless."

Von Neumann was quick to adapt, forgoing his work on the foundations of mathematics to explore applied realms. He claimed to have never read another paper in symbolic logic again. It was doubtless difficult to lose the foundation on which he had built his career. "I know myself how humiliatingly easily my own views regarding the absolute mathematical truth changed during this episode, and how they changed three times in succession," he'd later lament. He abandoned his dream of building an unassailable fortress for mathematics and traded it for the hope of building "synthetic rationality"—now known as artificial intelligence.

Around the same time, it had become clear that von Neumann could no longer stay in Europe. Antisemitism had been on the rise for decades, and tensions were particularly high in Hungary. Hungary's Jewish population had grown dramatically throughout the nineteenth century, more than doubling as Jews fled Russian pogroms in search of religious community. In 1919, Hungary was rocked by a short-lived and disastrously inept Soviet Republic led by Béla Kun, a Jewish journalist. In its aftermath, counterrevolutionary forces targeted Jews for public hangings and torture, and a modern Hungarian identity cohered around antisemitism. In 1920—the year the German Nazi Party was founded—Hungary passed a law limiting Jewish rights and restricting Jews' access to higher education. Threaded through von Neumann's childhood was the constant threat of extinction. As a child, he'd seen Jewish families flood into his city to escape anti-Jewish riots and violence in the east, only to face discriminatory laws curbing their job opportunities and educational prospects. His response was to bring his mathematical genius to bear on the problem of human rationality and illuminate the hidden forces of history. With games, he sought to survey the shape of a person's mind based on the simplest data possible: the choices they make in the world.

The threat of war radiated with growing intensity in the 1930s, veiled by a misplaced faith that European civility would avert bloodshed. In 1933, the year Hitler came to power, von Neumann was offered a full-time position at Princeton's Institute for Advanced Studies and leaped at the opportunity to leave. Von Neumann brought a European sensibility to the sleepy college town: a fussy aesthete, he dressed like an old-world banker in three-piece suits. He loved entertaining guests at his alcohol-soaked parties by drawing from a prodigious mental catalog of jokes. He had, according to colleague Raymond Seeger, "some arrested emotional development, such as an insensitivity to the feelings of women and a lack of sentiment." He'd also fallen into despair. He reflected on his disillusionment during the period leading up to World War II in a letter to his wife written years later:

> I feel the opposite of a nostalgia for Europe, because every corner *I knew* reminds me of the world, of the society, of the excitingly nebulous expectations of my childhood [. . .] of a world which is gone, and the ruins of which are no solace [. . .]. My second reason for disliking Europe is the memory of my total disillusionment in human decency between 1933 and September 1938, the advent of Nazism and the reaction of humanity to it—in that period I suffered my life's greatest emotional shock.

By 1935, Hitler had announced his plan to rebuild Germany's army, a program that had already begun, in secret, on Kriegsspiel boards in military clubs. It would take Germany's 1939 invasion of Poland to force other European nations to finally confront his threat. The Hungarian physicist Rudolf Ortvay was an old family friend and regular correspondent of von Neumann's, and he relayed news from Budapest to the anxious expat. European citizens, he wrote, were falling prey to an "excessive cult of the will." Both saw war in Europe as inevitable. Von Neumann wrote to his friend that "the whole affair is a pathological process and, viewed clinically, is a plausible stage of further development. It is 'necessary' even emotionally—if it is permissible to use the word 'necessary' in this connection." These pressures had to find release. Europe, von Neumann wrote, would be justifiably unseated from its dominant position in world affairs and its moral and intellectual weight diminished.

Starting in 1938, Hungary passed a series of increasingly restrictive antisemitic laws, stripping Jews of equal citizenship status. That same year, von Neumann's first wife left him, having fallen in love with a physics postdoc at Princeton. It took von Neumann little time to convince a woman he'd met on a previous visit to Hungary, Klára Dán, to divorce her husband in Budapest and move to Princeton with him. Despite the growing Nazi threat, von Neumann's and Dán's families remained reluctant to leave Hungary, paralyzed by an unrequited devotion to their homeland. The pair finally convinced their families to come to America in 1939. Dán's father died by suicide shortly thereafter. The atmosphere was grim: in a letter to von Neumann not much later, Ortvay reported that Leo Libermann, a professor well known to both, had also taken his own life. "In the state of the world, one cannot find great joy," he concluded. "I see it as slipping downward."

To distract his new wife from her fresh sorrow, von Neumann

invited her on a trip to the University of Washington, where he planned
to give several lectures on games. He had been revisiting his earlier
work, perhaps inspired to model the rising tensions in Europe. Could
mathematics prescribe or predict the best moves in complex conflicts?
Whereas Kriegsspiel allowed military strategists to zoom out on a
particular conflict, von Neumann's work on games would give strategists the tools needed to express *any* conflict. He'd been expanding
his two-player results to games with more players. In multiplayer
games, players can form coalitions, banding together to beat other
players. This mirrored what was happening in European politics, as
countries began to align and ally themselves along the political spectrum. The week after his Washington trip, von Neumann wrote to
Ortvay: "Unproductive as it is to meditate upon political problems, it
is hard to resist doing so. [. . .] In particular, it appears in all likelihood
that not two, but three or four, enemies are facing one another." However, he'd also discovered that his math could not predict which players would coordinate with one another. This was dictated, instead, by
external social pressures not contained in the game rules. "Which
'winning' coalition is actually formed," he'd write in an unpublished
manuscript, "will be due to causes entirely outside the limits of our
present discussion."

Historian Robert Leonard argues that von Neumann's interests
reflected a defiant desire to discern lawfulness in human affairs during
a time of terrifying disorder. Von Neumann's axiomatizing impulse
ran to his core. He professed himself "horrified" of messy fields like
biochemistry: "I cannot accept that a theory of prime importance,
which describes processes which everybody believes to be elementary,
can be right if it is too complicated, i.e., if it describes these elementary processes as being horribly complex and sophisticated ones."

Game theory was, in short, von Neumann's attempt to discern rational human motivations in the senselessness of European politics, to

chart the "economic pressures" driving human behaviors and predict where they might lead. He had established his treatment of games on the premise that players—people—are fundamentally rational, orderly, and subject to quantifiable forces, even if those forces drive them to counterintuitive ends. Von Neumann's letters to Ortvay emphasized ideas of equilibrium, simplicity, selfishness, and the centrality of social norms as stabilizing factors. Ortvay, possessing a front-row seat to the tensions at play, did not agree:

> I believe that these are economic forces only to a very slight degree; rather, they are enormously primitive and brutal passions, and the 'economic' reasons are in many cases only suitable for the purpose of letting modern man hide the real reasons from himself.

This did not dissuade von Neumann from his quest to discover the physics of human nature. He was not alone: other scholars also sought to axiomatize the social sciences around the same time. The Viennese mathematician Karl Menger had attempted to develop a mathematical foundation for morality in his 1934 book, *Morality, Decision and Social Organization: Towards a Logic of Ethics*. Menger, in turn, influenced Oskar Morgenstern, a young Austrian economist. As he was endlessly pleased to divulge in gossip, Morgenstern was the son of Frederick III's illegitimate daughter. He was also therefore the great-grandson of the same Kaiser Wilhelm who'd forged a new German Empire thanks, in part, to Kriegsspiel. Inspired by Menger's ideas, Morgenstern hoped to systematize economics. The existing literature was a mess. Most of his work in the 1930s amounted to criticism of his field. Economists were working with inconsistent definitions of the same quantities and predicating their work on impossible assumptions, either explicitly or implicitly through sketchy math.

Economists had borrowed the notion of equilibrium from physicists, who used it to describe systems at rest. A boat's depth in the sea reflects the equilibrium point of a battle between gravity and the buoyant force of water. Economic equilibrium models were in vogue, but Morgenstern was skeptical of these approaches because they're lifeless. A static model is a terrible approximation of a real market. In his view, an economic model that can't say anything about what happens when something changes "hardly deserve[s] the names of theory and science." Furthermore, equilibrium models required market participants to have perfect foresight. Morgenstern wondered: "The foresight of whom? Of what kind of matters or events? For what local relationships? For what period of time?" What could foresight even mean when reasoning about other *people*? How could they predict one another without falling into an infinite loop? *I know that you know that I know that you know . . .*

Morgenstern was fond of using a scene from a Sherlock Holmes mystery to illustrate his point: Sherlock, attempting to evade his archenemy, Moriarty, boards a train to Dover. Unfortunately, Holmes sees Moriarty arrive just in time to glimpse Holmes as his train leaves the station. Holmes must decide whether to ride the train to its final destination or disembark along the way. He must make that decision based on where he thinks Moriarty will head first, which depends on what Moriarty thinks Holmes will do, and so on, ad infinitum.

The problem was rooted in an issue Karl Marx had identified almost seventy years prior. Man-in-isolation thought experiments were a standard of moral philosophy, and economists had happily adopted the tradition. "Robinson Crusoe's experiences," Marx notes, "are a favourite theme with political economists." Crusoe, the hero of Daniel Defoe's famous novel, is a shipwrecked traveler who, stranded on an island for decades before being rescued, becomes self-sustaining, creating all the objects he needs to live. Crusoe looked, to economists, like the irreducible atom of a market, and they developed their models based on this idea. However, they ignored interactions between the individuals making up the market, as though the market were a cloud of isolated Robinson Crusoes rather than a network of interrelating subjects. Marx argues that this led to economic models that were inherently objectifying, excluding humans as subjects and instead modeling an economy of objects: collections of navel-gazing, noninteracting individuals.

Beyond his colleagues' sloppiness, Morgenstern objected to how economists framed their questions. Most economic theories were little more than price predictions supposing some institutional structure: *Assuming a market with perfect competition, determine the price of wood given x supply and y demand.* But how did this institutional structure arise in the first place? Rather than assuming an institution as given and predicting a corresponding price, economists should contend with a more interesting question: *Given a set of rules, what institutions and price pairings are possible?* Economists should do better than merely describing an existing dynamic; they should work to build tools that can express *any* possible dynamic. Economics, in short, was a fragmented field whose practitioners were asking the wrong questions with the wrong tools, offering useless and static models built on a foundation of impossible assumptions.

In 1938, the year that Hitler annexed Austria, Morgenstern left his post in Vienna for Princeton's Institute for Advanced Studies. It didn't take long for Morgenstern and von Neumann to discover their mutual interest in formalizing human interactions—von Neumann through the lens of games and Morgenstern through economics. They merged their efforts and extended von Neumann's earlier work into the now canonical text *Theory of Games and Economic Behavior*, published in 1944. In the book, they describe how a collection of interacting players, idealized under certain assumptions and restrictions, will strategically behave to maximize their payoffs. "We hope to establish satisfactorily [. . .] that the typical problems of economic behavior become strictly identical with the mathematical notions of suitable games of strategy." At the time, many scholars believed that economics would permanently resist mathematical formalization. Clarified through games, economics began its transmutation into a more rigorous discipline—cosmetically, at least.

The authors compare the study of games to Newton's apple. Just as the trajectory of an apple falling to earth reveals laws of physics explaining motion on planetary scales, games can be considered minimal models from which more complex dynamics can be constructed. Whereas Newtonian mechanics describes how entities with mass affect one another, game theory describes how entities with desires interact. Using games as a foundation, researchers can reason beyond the mechanics of passive matter to the mechanics of active agents with preferences and goals. Just as gravity draws an apple to earth, self-interest draws players into predetermined strategies that allow them to do as well as mathematically possible. Von Neumann was convinced that his model accurately captured the reality of human behavior. His childhood friend and colleague Eugene Wigner claimed that this worldview colored his moral judgments, quoting von Neumann as saying: "It is just as foolish to complain that people are selfish and treacherous as it is to complain that the magnetic field does not increase unless the electric field has a curl. Both are laws of nature."

Von Neumann and Morgenstern give concrete definitions for notions economists had previously only loosely construed. The concept of utility had been introduced in the eighteenth century, commonly attributed to the philosophers Jeremy Bentham and John Stuart Mill, though it had antecedents in earlier work. Utility is meant to measure the pleasure or advantage a person gets from something. But this isn't easy to quantify—it's tough, for example, to define in rigorous terms the pleasure different consumers might get from the same object.

People have not only different preferences but also very different base-lines. The maximum pleasure one person experiences could theoretically be lower than the minimal pleasure of someone else.

In 1937, the economist Paul Samuelson worked out a definition of utility that could be measured in purely observable terms, without referring to hidden psychological factors. His insight, now known as revealed preference theory, was based on Daniel Bernoulli's insight that agents act to maximize their payoffs, or utility. Through its choices, an agent reveals how much it values the outcome it's pursuing; that is, the outcome's utility. A thing's utility is measured by how consistently the agent chooses that thing. An agent is an agent because they act to maximize utility. They choose all actions in a bid to maximize utility, utility being that which they maximize. Though circular, this technically fixed the problem. As mathematician Ken Binmore puts it:

> The modern [utility] theory doesn't say that Eve chooses a rather than b because the utility of a exceeds that of b. On the contrary, the utility of a is chosen to be greater than the utility of b because it has been observed that Eve always chooses a rather than b.

Von Neumann formalized Samuelson's insight and explicitly incorporated the notion of risk into a player's choices. All choices are like a gamble, and von Neumann modeled them as such. When we buy a stock, for instance, we don't know what it will be worth in the future. We estimate the probability that it will rise and buy it if the bet seems safe enough. We may still choose to make a wager with a low likelihood of success if the outcome is something we want enough—like putting in the effort to apply for a highly competitive job, even if it's unlikely we'll get it. In this framework, a rational actor will always make choices that they believe will maximize their expected utility, balancing their desire for a given option with its risks. Given certain assumptions, their preferences can be described by a utility function, which can be used to predict their strategies and choices. Von Neumann's elegant formulation transmuted subjective individual desire into a shared frame of reference, defined entirely by an agent's actions in a collective playing field. He derived all of this in the span of a single afternoon. Morgenstern later wrote in his diary that watching von Neumann solve in one day a problem that had plagued economists for years "gave me great satisfaction, and moved me so much that afterwards I could not think about anything else."

Game theory is predicated on the assumption that players are rational. The game-theoretic definition of "rational" has evolved over the years—von Neumann and Morgenstern's conception was impossibly restrictive, requiring that players possess infinite foresight and computing capabilities in order to calculate their optimal choices. The technical definition has nothing to do with the players being logical or unemotional or even intelligent. In its most general form, it's defined by the choices players make. Rational players make choices that maximize their payoffs, or utility. This requires that they have consistent, well-defined preferences. For example, if a player likes apples more than oranges and oranges more than bananas, they should prefer apples over bananas. A rational agent should be able to rank their preferences for given outcomes, reason about which actions will lead to which outcomes, and make choices that lead to their preferred results. If these criteria are satisfied, an agent is said to be rational, and an economist can predict the agent's choices so long as they know the agent's preferences.

As the Nobel Prize–winning economist* Amartya Sen points out in his classic paper "Rational Fools," we would hardly consider this "ratio-

---

* What is commonly called the Nobel Prize in Economics is officially the Sveriges Riksbank Prize in Economic Sciences in Memory of Alfred Nobel.

nal" agent, based on the consistency of their choices, meaningfully intelligent. Intelligent choices don't fall into a single fixed ordering; they depend on context. A person might prefer a parka to a tank top in the winter but probably not when visiting a tropical island. They might prefer a toothbrush to a bulletproof vest at bedtime but not in the depths of combat. One could even argue that game theory's fundamental assumption—that players have ranked, static preferences—is invalidated by the existence of games themselves. Players in the heat of a Monopoly game covet nothing more than its pastel paper money, which becomes instantly worthless as soon as the game ends. During games, we temporarily suspend our true desires to submit to a fictional play world in which there is nothing we want more than to pack tiny plastic houses onto a particular square of cardboard real estate.

Unlike "Robinson Crusoe" economics, game theory provided the framework needed to describe interactions. Morgenstern hoped it might become a tool for economists to study the emergence of institutions, or "orders of society," given social rules. Yet *Theory of Games and Economic Behavior* addressed few of Morgenstern's reservations about the state of economic theory. "One of the ironies of the situation was the extent to which Morgenstern's original concerns were trampled and mangled," writes historian Philip Mirowski. Morgenstern had spent his career criticizing economists' unrealistic models, then established a field that modeled agents with perfect foresight. Game theory is also static. Von Neumann and Morgenstern admitted that "a dynamic theory would unquestionably be more complete and therefore preferable." But they pointed to other sciences as evidence that it would be futile to build a dynamic theory before understanding the static case. Morgenstern had abandoned his lofty ambitions, noting in his diary: "Johnny says we should wait 300 years for an applicable dynamic theory or 100 years if one is exceptionally optimistic."

Some of these restrictions were relieved over time. For instance, the political scientist Herbert A. Simon developed "bounded rational-ity," recognizing that people have finite processing capacity and often make decisions that aren't optimal but good enough. Perhaps the big-gest problem with the new theory was that it dealt only with zero-sum games. This made it a particularly poor tool for economists. Markets are supposed to be positive sum; their very purpose is to generate value. But the restriction to zero-sum games was a handy mathemati-cal trick, acting something like a conservation law in physics. Because players' interests were balanced against one another, the minimax so-lution emerged as an equilibrium point and gave theorists something to grab on to in players' slippery psychological depths. Players in a two-person zero-sum game never have to devolve to that infinite re-gress of "I think that you think that I think . . . ," because the minimax strategy holds regardless of what the other player does. "In a zero-sum two-person game," write von Neumann and Morgenstern, "the ratio-nality of the opponent can be assumed, because the irrationality of his opponent can never harm a player." A player can't do worse than the minimax, regardless of their opponent's strategy, so all players should therefore gravitate to the minimax strategy.

Von Neumann was less concerned than Morgenstern with provid-ing economic interpretations of their models, though the final text reads as almost apologetic for their failure to furnish a more realistic account. Morgenstern predicted that it would take a long time for game theory to be embraced by economists. He was right: economists dutifully ignored *Theory of Games and Economic Behavior* for decades, finding it too narrow to solve anything of interest. The book also an-tagonized many economists because its authors both criticized their field and introduced an alien form of mathematics. Game theory was primarily advanced by mathematicians, not economists, during the early decades of its existence. It remained relegated to something of a sideshow until gradually, over the course of half a century, it dragged the field of economics into itself.

By 1950, Princeton was the world epicenter of game theory. This "Camelot" was based in Princeton's math—not economics—department. The economist Martin Shubik, a student of von Neumann and Morgenstern, recounted:

> the contrast of attitudes between the economics department and the mathematics department was stamped on my mind soon after arriving at Princeton. The former projected an atmosphere of dull business-as-usual conservatism of a middle league conventional PhD factory [. . .]. The latter was electric with ideas and the sheer joy of the hunt.

Games including Go and von Neumann's childhood favorite, Kriegsspiel, were wildly popular among the faculty and students, and competition ran heavy in the department's common room. Amid these players was John Nash, a precocious young graduate student who had been tersely recommended to Princeton by his undergraduate adviser: "Mr. Nash is nineteen years old and is graduating from Carnegie Tech in June. He is a mathematical genius." In a one-page paper published in 1950, Nash revitalized economic thought.

Von Neumann and Morgenstern's analyses were primarily limited to two-player zero-sum games. These have a clear equilibrium point: players gravitate toward minimax strategies. After they added a third player, though, things got weird. Two players might band together against the third, then split the payoff. But game theory can't predict how these coalitions form, which players will be in them, or what strategies they'll gravitate toward. Nash proposed a much broader definition of a game's equilibrium, one that generalizes to games with any number of players and any payoff structure, whether zero, positive, or negative sum.

In a Nash equilibrium, a player's strategy is the best response to the choices of all other players. The Nash equilibrium has the property that if every player announces their strategies simultaneously, no one will want to change theirs. Consider, for example, the side of the road we drive on. Even if no traffic laws enforce one side or the other, it behooves all drivers to pick a side to drive on and stick with it. There is a natural pull to a strategy of driving on either the left or the right. One wouldn't benefit from driving on the other side of the road unless everyone else agreed to do the same.

Nash's equilibrium is related to a result derived by the nineteenth-century economist Antoine Augustin Cournot in describing competition between a duopoly of rival companies selling spring water. "The state of equilibrium [. . .] is therefore *stable*; i.e., if either of the producers, misled as to his true interest, leaves it temporarily, he will be brought back to it by a series of reactions." Morgenstern must have been familiar with the result but either missed the connection or, like von Neumann, thought it an uninteresting result. Nash later reported that when he presented his work to von Neumann, von Neumann replied, "That's trivial, you know. That's just a fixed-point theorem." But Nash's equilibrium concept gave game theory the generality it needed to become more useful to the social sciences.

Von Neumann may have initially dismissed the result because he thought competitive games were not a good model of the real world. In reality—politics, markets, social circles—players almost always seek to form coalitions, yielding many possible outcomes. Nash published his results in a brief paper in 1950. This work—alongside his treatment of bargaining games, wherein cooperative players bargain to divide goods—earned him a Nobel Prize in Economics in 1994. By then,

Nash had been incapacitated for years by paranoid schizophrenia, as portrayed in Sylvia Nasar's *A Beautiful Mind*. Voices crowned him emperor of Antarctica, relaying messages from other worlds encrypted in newspaper articles. Despite the depth of his mathematical insight, he despaired that he'd never been able to live up to his full potential due to his uncontrolled illness.

The philosophies underpinning equilibrium economics were adopted and popularized by conservative, laissez-faire thinkers who used these models to rationalize pro-business policies. It's unfortunately common for economists to mistake mathematical equilibria for reality. Regulations aren't necessary, some argue, because people spontaneously gravitate toward equilibrium strategies. A famous "law" of economics has it that a perfectly competitive, unregulated market operating without constraint yields the greatest social welfare. But this is because scholars rebranded "economic surplus" as "social welfare." Unregulated markets don't guarantee that surplus is fairly distributed, so calling it welfare is ludicrous, unless we mean the welfare of the wealthiest.

To put it plainly: markets are not magical. But it is hard to argue with the willful naivete of a theorist whose model agrees with their ideology. Free-market fundamentalists contend that markets offer a kind of ethical alchemy. Selfishness is held up like a philosopher's stone, capable of transmuting what was once considered a sin into economic virtue. Selfishness not only drives productivity, it renders people predictable—a virtue to economists hoping to explain social dynamics. To restrict the market's operation in any way is not just bad business—it's immoral. The true followers of the laissez-faire religion claim that unregulated markets can solve any problem because markets are most efficient when perfectly competitive. But this theoretically perfect competition also requires impossibly narrow and unrealistic conditions. Libertarians, who tend to present their worldview as hard-nosed realism, have constructed a philosophy based on idealized math with no touchpoint to the real world so beloved by them.

The ascendancy of free-market fundamentalism owes more to PR than to academic rigor, as Naomi Oreskes and Erik M. Conway document in their book *The Big Myth*. In the 1950s, General Electric Corporation revived the flagging career of Ronald Reagan, then a New Deal Democrat, when they hired him to host the TV show *General Electric Theater*. The program preached the gospel of individualism and free enterprise, spreading GE's staunchly anti-union and anti-government ideology. Similarly minded corporations funded think tanks and lobbying groups to spread propaganda that framed market regulations as against freedom, democracy, and American values. They distributed millions of free copies of books by famous conservatives, including Friedrich Hayek, Milton Friedman, and Ayn Rand, who embraced game theory's core assumption of human selfishness. Game theory became a favorite contrivance in the neoliberal tool kit, accelerating the world's mass privatization.

To date, fifteen Nobel Prizes in Economics have been awarded to game theorists for topics ranging from regulating monopolies to games of asymmetrical information, like labor, insurance, and used-car markets. Game theory's logic has undergirded the design of novel platforms, auctions, and economic systems—from the algorithm that matches US medical students to their ideal residencies, to billion-dollar auctions of airwave space. Companies compete in game-theoretic auctions for search-engine ad space. Researchers use game theory to predict supply chain flows and organize communication protocols for connected devices. Through the power of game-theoretic market design and auctions, online goods and services have become so efficient that they've driven the demise of brick-and-mortar stores. Game-theoretic matching algorithms work to pair travelers with their ideal flights, job seekers with new careers, and socially alienated people with fanatic propaganda. Game theory's insights eventually bled from academic thought experiments to inform widely accepted "common-sense" beliefs. Game theory has shaped military strategy, dictated the designs of modern social and economic systems, and influenced how we see ourselves—for better and for worse.

6

The Clothes Have No Emperor

What makes you so sure that mathematical logic corresponds to the way we think? You are suffering from what the French call a "déformation professionnelle." Look at that bridge over there. It was built following logical principles. Suppose that a contradiction were to be found in set theory. Do you honestly believe that the bridge might then fall down?

STANISLAW ULAM

Researchers have put a notable population of humans through decades of psychological tests. Many of these subjects repeatedly deny the experimenters' requests that they share collectively held assets with other test subjects. When gifted an amount of perishable food wildly exceeding what any one person could eat, the majority of these subjects offer less than 10 percent of their hoard to another subject given nothing. Though they are able to deliver food to both themselves and other subjects at zero cost, these subjects as often as not ignore the other participants' requests for food. In many game-theoretic scenarios, researchers have determined this population to be the most selfish human subgroup known to science.

These subjects are toddlers. Children are more prosocial by age six and become more patient and reasonable bargainers. In many ways, real people have defied game theory's rational, self-interested assumptions. Even two-year-olds spontaneously share sometimes.

The players in game theory are not people, of course, but abstractions. Game theory doesn't make predictions about human behaviors. Based, as it is, on mathematical tautology, it doesn't have direct access to the real world. Game theory describes mathematical models whose "choices" are just the logical consequences of their utility functions. A game-theoretic agent is nothing more than an equation, a bundle of preferences destined to take actions in pursuit of these preferences. When game theory's predictions fail to match observed human behavior, this doesn't disprove game theory. Game theory cannot be refuted by empirical observations, as it's not empirically based. Von Neumann and Morgenstern are clear about this in their book. Mathematician Ken Binmore writes: "Mathematical theorems are tautologies. They cannot be false because they do not say anything substantive. They merely spell out the implications of how things have been defined. The basic propositions of game theory have precisely the same character." This does not make game theory uninteresting or useless, but we should be wary of accepting its lessons into the realm of established common sense.

Von Neumann and Morgenstern chose game theory's original assumptions not because they were accurate but because they made for simpler math, especially in a precomputer world. As their advisee Martin Shubik explains: "Economic man, operations research man and the game theory player were all gross simplifications. [. . .] Reality was placed on a bed of Procrustes to enable us to utilize the mathematical techniques available." Despite these limitations, theorists jumped on the opportunity to generate testable predictions of human behaviors, hoping that game theory would become the backbone of a more quantifiable social science.

The possibility of knowing how people will act before they themselves know seeded a fantasy too alluring to ignore. Game theory went on to have an outsize influence on US policy. The year 1948 saw the establishment of the RAND (Research ANd Development) think tank, primarily dedicated to military, economic, and political applications of game theory. It was spearheaded by Douglas Aircraft—a major military supplier facing massive cutbacks after the war—and funded by several federal agencies, including the US Army, Air Force, and Department of Defense. RAND analysts informed policy decisions ranging from the space race to health care. Game theory provided their principal tool set.

The most influential work to come out of RAND—and game theory's most famous thought experiment—is known as the prisoner's dilemma, developed around 1950 by Merrill Flood and Melvin Dresher. The prisoner's dilemma describes a class of games in which players, acting out of self-interest, harm all players, including themselves. Imagine that two criminals bearing shared responsibility for a crime are caught and interrogated separately. If they collude and both stay silent, each will only serve one year in prison. If they mutually betray one another, each will serve two years. And if one betrays the other but the other remains silent, the betrayer will be set free, while the betrayed will serve three years in prison. The point here is that it's globally optimal for both parties to stay silent. But according to game theory, both criminals will always betray one another to try to secure their individual, immediate freedom. In so doing, both are doomed to serve two years, though they could have cooperated and served only one year.

Defection is the Nash equilibrium of any game structured in this manner. Yet the players fare worse by acting selfishly than they would have had they acted in a socially responsible way. They betray their self-interest by seeking their self-interest, undermining Adam Smith's insight that acting selfishly can paradoxically benefit everyone. As Flood and Dresher discovered, this isn't always so. Pierre-Simon Laplace claimed that "probability theory is nothing but common sense reduced to calculus." But this strange logic conflicted with common sense. When they first reported their paradox, Flood and Dresher hoped that others would be able to solve it, but the prisoner's dilemma has remained a thorny enigma at the heart of game theory.

Flood and Dresher tested their colleagues in a mock scenario and quickly found that real people don't act like their game-theoretic counterparts. A purely rational agent would defect and rat on their partner. Instead, players were more apt to collude and so performed better than the pessimistic "rational" prediction. When Flood and Dresher put two colleagues in a repeated prisoner's dilemma game, one persistently cooperated and was taken aback when the other player didn't do the same. "The stinker," the habitually cooperative player noted of his defecting opponent. "He's a shady character and doesn't realize we are playing a third party, not each other. [. . .] A shiftless individual—opportunist, knave." He used his responses to reward and punish the other player's choices. "This is like toilet training a child— you have to be very patient." Over time, the defector became more cooperative.

The prisoner's dilemma became ubiquitous in empirical psychology research: political scientist Robert Axelrod dubbed it "the *E. coli* of the social sciences," after microbiology's favorite model organism. Researchers have used it as a lens to understand why humans don't act like their rational game-theoretic projections. People, economists and psychologists have concluded, are not very rational. Vernon Smith and Daniel Kahneman shared a Nobel Prize in 2002 for their decades of work uncovering a jungle of cognitive biases, such as temporal discounting and regret aversion. Kahneman carefully details about a dozen of these in his opus, *Thinking Fast and Slow*, though researchers have now documented hundreds of biases. The field continued to grow into what's now known as behavioral economics, the study of human decision-making: a mishmash of psychology and economics. It's a challenging field that requires heavy caveats—human decisions are notoriously messy and hard to characterize. Though experimental protocols for behavioral research have improved over the years, psychology studies are often limited by low participant numbers, artificial conditions, and subjects representing an unusually narrow population: young white college students. Behavioral economics also comes primed with its own biases, in part because it is richly funded by corporations seeking scientific confirmation of their operating assumptions. As such, it's one of the least reputable fields of science, and researchers have failed to replicate many of its classic findings. Despite these caveats, the bulk of evidence indicates that humans are not purely rational actors. However, researchers have put more effort into cataloging cognitive biases than into understanding why they might exist. What's causing these discrepancies, and what, if anything, can they tell us about human nature?

Before we dig into the ways people aren't rational, it's important to unpack what it means to be "rational" in this context. At a 1966 conference, political theorist Thomas Schelling petitioned his colleagues to use a less loaded term. While the lay definition connotes "logical," in the context of game theory, a rational agent simply acts in alignment with its goals, maximizing its expected return. This doesn't mean that rational behaviors, in the game-theoretic sense, are intelligent, or preferable. In game theory, players don't need minds at all, only desires. Rational choices are those made in pursuit of these desires. Given a player's preferences, or utility function, game theory reveals a player's optimal decisions without ever having to ask the player for input. A player's preferences might be entirely "irrational," by the popular definition: a masochistic agent, for instance, might seek its own suffering. For that agent, rationality would mean making choices consistent with self-harm. A rational agent's goal can be selfish or unselfish—they may want to win money for themselves or give it all to charity. Their goal can be reasonable or unreasonable, self-preserving or suicidal. A rational agent simply makes choices that maximize its chance of winning its aim. Modern media is saturated with reports on the many failures of human rationality, but these smuggle a value judgment into a technical definition. If the mathematical property we now know as rationality had instead been given a label with a negative valence—"greedy maximization," say—headlines would read, somewhat more charitably, "People Are Not Greedy Maximizers" instead of "People Are Not Rational."

Game theory is a field of pure math and, as such, can't make empirical predictions. Nevertheless, this once niche mathematical discipline has imperialized many social sciences, including economics, political science, law, ethics, and psychology. Though economists have labored for years to connect economic theory with more realistic human behaviors, economics textbooks often feature ideas markedly out of date. Physicists, the economist Herbert Gintis points out, regularly update their models to accord with experimental data. The deeply unintuitive field of quantum mechanics was developed over decades, initially spurred by an experimental anomaly involving blackbody radiation. Theorists took every new experimental upset to heart: each indicated that there was something more to understand. A succession of updated models eventually led to the conceptual breakthrough of quantum mechanics. The same has not been true for economics. Surveying a classic microeconomics textbook, Gintis concludes that the text, "despite its beauty, did not contain a single fact in the whole thousand-page volume." Such texts serenely herald market "laws," when what's meant is "tautologies." Their authors build mathematical castles in the sky, untethered from reality.

The biggest problem, Gintis argues, is the assumption that people act out of self-interest. There's too much evidence to the contrary.

Many psychology studies have indicated that people are loath to lie to one another, even at personal cost. People are especially honest when substantial money is on the line. Subjects take cooperation so seriously that they punish other players for not cooperating, regardless of whether it affects them personally. Institutional and social mechanisms for punishment and reward (like prison, fines, and gossip, or trade agreements, dividends, and reputation) may have emerged as tools to enforce cooperation. It's little wonder that a species that owes so much of its success to large-scale collaboration would exhibit other-focused behaviors. Mathematicians, Gintis contends, need to work on updating game theory's premises to better account for that.

Another difficulty in reconciling game theory with behavioral studies is that game theory is all about preferences, and experimentalists can't always tell what players really want. In fact, players don't have preferences—preferences define the player, and the game. There is no person making choices; choices are logical conclusions of the player's preferences, and those preferences are measured, tautologically, by the choices the player makes. Sometimes the emperor has no clothes, but in game theory, the clothes have no emperor.

Knowing an individual's preferences means knowing what they will choose in every conceivable situation. Rational-choice policies threaten to undermine democratic ideals, as experts substitute a theoretical "best" choice for people's actual choices. In game theory, writes historian S. M. Amadae, "consent is rendered superfluous." Game theory was intended to be a mathematical description of agency yet divests players of it.

In a laboratory environment, a player's choices are rigidly prescribed and don't necessarily reflect the dynamics of the real world. In this artificial scenario, preferences are imposed rather than measured: experimenters create a game with a particular payoff structure and expect players to act according to that structure. But research subjects may be playing a very different game from the one researchers think they've designed. Their preferences may include intangible rewards—like valuing cooperation—that can't be easily quantified. For simplicity's sake, theorists often leave such considerations out, instead assigning money or points as the sole measure of utility. But this doesn't mean that players don't value what can't be measured.

Flood and Dresher set up their prisoner's dilemma experiment with the assumption that players cared only about minimizing their fictional prison time. Real people might be more concerned with other things. Their cooperating subject, after all, believed that he and his opponent were pitted against a third party. Two participants who don't know or care about one another might behave like cynically defecting prisoners, driven to serve the least prison time. But if we replace these players with two star-crossed lovers, who value the other's freedom more than their own, the prisoner's dilemma is transformed into a lover's delight. These players' payoffs will differ from those of our agnostic prisoners, and their equilibrium strategy will be to cooperate.

Rationality, as discussed, means that a player makes choices consistent with their desires. The desires themselves needn't be rational—in the everyday sense—or self-serving. A self-interested choice can still be unselfish. Rather than interpreting a player's cooperative bias as a "failure" of rationality, perhaps we should take their choices at face value. Maybe they *were* acting out of self-interest—they just cared about something the researchers hadn't accounted for. The player may have preferred to cooperate for fun rather than to maximize meaningless points or minimize costs like fake jail time. A player's preferences, not our assumptions about them, determine the games they actually play.

In game theory, players' preferences are usually assumed to be fixed. But psychologists have found that people's choices are sensitive to a game's context. Players behave differently in a prisoner's dilemma depending on how it's presented to them. If it's called the Wall Street Game, players are more likely to defect. If it's called the Community Game, they're more likely to cooperate. The two games have the exact same structure, payoffs, and penalties (e.g., jail time), yet their titles alone cause players to act differently. Games are the science of *interdependent* interactions, and players may take cues from a game's framing narrative to predict whether other players will defect or cooperate themselves. Players might also use this context to anticipate any intangible payoffs, such as abiding by shared values and social norms. Gintis posits that norms act as a "choreographer," coordinating players' behaviors. Players take cues from the social context to predict what other players will do, then choose their own strategy accordingly.

Real people, unlike their classic game-theoretic counterparts, learn. A game teaches its players how to behave. The game Mafia rewards its players, often against their better nature, for lying to their friends. A player determined to win the video game *Civilization V* in a nonviolent manner will nevertheless find the game's ersatz Napoleon Bonaparte driving tanks across their borders. A game's structure, both explicit and implicit, incentivizes the players' behaviors. Though players initially tend to cooperate in a one-shot prisoner's dilemma with strangers, they defect more often over repeated exposure, as though learning the Nash equilibrium through experience. Players learn to defect over time rather than determining the strategy a priori. It's a race to the bottom: the more their counterpart defects, the more they themselves will defect. This doesn't mean that humans are, by nature, irredeemably selfish; their choices simply reflect the game's incentive structure. "It is the *logic* of the prisoners' situation, not their psychology, that traps them in the inefficient outcome," writes philosopher Don Ross. It's perfectly possible, Gintis argues, that over the course of evolution, humans internalized cooperative norms to overcome the inefficiencies of prisoner's dilemmas, and that's why their first instinct is usually to cooperate.

This is an important consideration. In game theory, a player's preferences can be divined by the choices they make. In reality, many of us make choices out of necessity, not desire. We've learned to make choices rewarded by the games we play—games we're sometimes forced to play. The problem with defining people by their choices is that these can be coerced. People are often rewarded for making choices that don't reflect their true preferences but align with social or corporate interests. In countries where wages have not kept up with the cost of living, people might "choose" an exploitative workplace because it's their only option. Pare a person's options down enough, push them to the brink of subsistence, and see whether they can still honor their true values. Buried in the assumptions of game theory is the notion that all value can be objectively measured and included in one's model. In theory, the measure of utility can and should include a person's moral values. Practically speaking, economic welfare is most often measured in terms like GDP, wages, or savings. It becomes a system insensitive to everything except what can be quantified in dollars. It cannot, for instance, register human misery, save what can be profited by it.

It is a false equivalence to claim that someone's preferences are what they've been forced to choose. An HR employee might protect a sexual predator because their paycheck—and their family's health insurance—depends on it. A financial adviser might prefer life as a kindergarten teacher but wouldn't be able to afford rent. People may be compelled, by economic or social pressures, to make choices that don't reflect their true preferences, then are told that these forced choices mean something about human nature. "The market," Binmore writes, "is therefore the final step in a process that first leaches out the moral content of a culture and then erodes the autonomy of its citizens by shaping their personal preferences." Increasingly, economists and engineers use game theory to design new social and economic structures. It's of utmost importance that we create robust systems that take this preference warping into account. Game theory posits humans as a fixed bundle of preferences, when they are, in fact, learning systems. This remains an important area of development, and there are branches of game theory that incorporate player learning. However, these still often treat preferences as fixed and not necessarily changeable or learnable themselves. It's common for people to confuse their personal goals with those quietly manufactured for them by the games in which they participate, wittingly or unwittingly.

The internet has revealed just how plastic our preferences are: people can be radicalized by consuming progressively extreme content. Because social media has been effectively gamified, people may be socially rewarded for adopting stances that garner them attention or approval from niche communities. They may adopt beliefs about the world not because they reflect reality but because they're reinforced by social rewards, like community status. Plainly false ideas can go on to have real social and political repercussions if they attract enough devotees—take QAnon, for example.

In addition to warping our preferences, games can shape our behaviors. Theorists have known for half a century that games can elicit totally irrational behaviors in their players. While still a graduate student at Princeton, Martin Shubik designed games that invoked "pathological phenomena" in their players. In his ingenious dollar auction, players bid to win one dollar. They can start bidding as low as five cents—a deal too good to miss. The catch is that the second-highest bidder will owe the auctioneer the amount of their losing bid. A bidding war begins as soon as a second player bids for the dollar. Say that player one starts off bidding five cents. If the second bidder bids ten cents, hoping to net ninety cents, this incentivizes the first bidder to bid fifteen cents, else lose the five they initially bid. This causes the second to bid twenty cents, and so on. This ratcheting up happens well past bids of a dollar: the game incentivizes participants to continue bidding because winning the dollar would mean they'd at least get to offset their losses. The position of both bidders only gets worse as they continue to throw good money after bad. The only way to win is not to play, yet players often take the bait. Shubik claims to have won thousands of dollars demonstrating this game.

Ariel Rubinstein is one of game theory's most accomplished practitioners, and one of its most vocal critics. A celebrated Israeli economist, he's considered by many to be overdue for a Nobel Prize for his foundational work on bargaining theory, among other contributions. His parents immigrated to Israel after losing most of their families to the Holocaust. Rubinstein's website features a massive database of "cafés where one can think"—he does his best work in coffee shops, despite loathing coffee. Game theory, he contends, is a beautiful field of pure mathematics, but it holds no lessons for the real world. In fact, he suspects it's made the world worse. In his afterword to von Neumann and Morgenstern's *Theory of Games and Economic Behavior*, Rubinstein writes:

> Personally, I am not sure that game theory "improves the world." Overall, economics, and game theory in particular, is not just a description of human behavior. When we teach game theory, we may be affecting the way people think and behave in economic and strategic interactions. Is it impossible that the study of game theoretical considerations in economics makes people more manipulative or more selfish?

In many academic circles, game theory has become a standard against which human behavior is measured. But because people learn, this yardstick can distort the system it was invented to assess. It all boils down, again, to human plasticity. Whereas naive participants in a game tend to use commonsense strategies, players who have had previous exposure to game theory—in a college course, for example—tend to use strategies that are theoretically correct but practically worse. They know the "right" answer—the Nash equilibrium solution—and so may choose the inefficient outcome, like defecting in a prisoner's dilemma. Because game theorists have deemed it the rational solution, educated players go against their instincts and choose the cynical strategy. Rubinstein calls these tragically misanthropic players "members of the Victims of Game Theory organization." He argues that in real life, "their achievements would be inferior to those who had not become wise by studying game theory. This does not prevent some strategic experts from treating the game theory solution [. . .] as a sacred rule." Those who thoughtlessly subscribe to the logic of game theory, Rubinstein suggests, are doomed to worse outcomes because of their faith. Similarly, Hermann Hesse writes in *The Glass Bead Game* of "doubts whose very existence or possibility you had only to know about and you instantly began to suffer them."

Rather than centering our study of human behavior on abstract theory, we can learn from the source of human behavior: the brain. Several classic biases discovered by psychologists—our "failures" of rationality—appear to be signatures of the dopamine system. In economics, utility represents the sum of all value taken from a good or service. In the brain, dopamine encodes predictions about rewards, including sex, money, food, water, and socialization. Dopamine looks, in short, like the biological correlate of economic utility. This may explain why people are more prosocial than their game-theoretic counterparts: several studies have shown that cooperation itself triggers dopamine release—that is, people naturally find cooperation rewarding. Unlike the rigid preferences of game theory's agents, dopamine neurons respond to rewards in a dynamic, context-dependent manner, suggesting, again, that humans are better modeled as learning systems than as static mathematical equations.

Dopamine tracks with several other famous biases as well. The brain's dopamine response also reflects the economic notion of marginal utility: the more of something one has, the less one wants it. A thousand dollars is more precious to someone with zero dollars than to someone with a million. Wolfram Schultz's group found that the dopamine neurons of a thirsty animal respond strongly to cues predicting a juice reward. As the animals become sated, however, that response weakens. The more juice they've had, the less they value it. This is the basis of the hedonic treadmill: people habituate to improvements in their standard of living, then seek more extreme luxury after they adapt to their new reality.

The same goes for the bias known as temporal discounting. People tend to value immediate rewards more than future ones, even if the future rewards are larger. Most people prefer to watch a movie than to study for a test, for instance. The payoff for doing well on the test may be larger—like securing a better job—but won't come for months or years. Schultz and his colleagues trained monkeys to press a lever after being cued by a light flash to collect a juice reward. In trained animals, the dopamine neurons fired in response to the light cue that predicted the juice. When the team added a second light flash before the original light cue, the dopamine neurons learned to respond to the earlier flash, albeit less strongly. The further in time the cue was from the actual reward, the weaker the dopamine (and therefore motivation) response, reflecting our tendency to mortgage the future for the present.

The dopamine response also relates beautifully to another cognitive tic: we are more sensitive to relative than absolute measures. Say that two research participants perform the same task for a fee. The first person is told that they will be paid five dollars for the task, and the second person is told that they will be paid fifty dollars. If the first person receives six dollars instead of the five they expected and the second receives forty-eight dollars instead of the fifty they expected, the first person will feel more pleasure than the second, even though the second person has made, in absolute terms, eight times more money than the first. Recall that dopamine neurons track reward prediction errors. The strength of a dopamine neuron's firing reveals, for instance, how much an actual reward differed from the animal's expectation. Dopamine neurons increase their firing if given more of a reward than expected and decrease their firing if given less than expected. They register relative differences between reality and belief, not the magnitude of the reward itself. What matters is the deviation from expectation.

What economists have misnamed biases aren't defects obscuring the lens of pure reason—they are the way thinking works. Learning-based theories, like reinforcement learning, explain many human behaviors and biases better than game theory. Yet this has not stopped academics from situating game theory at the nexus of the social sciences. Nor has it prevented popularizers from rebranding game theory's logic as business acumen. But humans aren't static equations. They modify their behaviors and learn from the context of the game they're playing (whether Wall Street or Community). Beliefs, likewise, play a role in the dopamine system's predictions. This has two important implications: First, our behaviors will be shaped by what games we're playing, so we'd better be careful about how we structure these games. Games can teach us to misbehave—not because we humans are, at core, bad, but because games may reward us for acting so. Second, regardless of the intended structure of the game we're in, we'll behave according to our beliefs about it, which can change the very nature of the game we're playing. Other players' choices can influence our beliefs. One player who repeatedly defects in a repeated game of prisoner's dilemma can incite other players to defect as well. Our beliefs and interdependent choices tangibly affect our realities, defining the games played at the societal level.

Game theory, pioneered by a traumatized man hoping to plumb the depths of human nature, has only ever been about math: the dynamics of a maximizing function, not of real people. We must be careful not to let its models mean anything about ourselves, and careful to prevent its fables from seeping into the domain of accepted wisdom. Let's consider the zero-sum bias. People across the globe tend to believe that all resources are limited, even when this is not the case. An isolationist might oppose immigration because they think that immigrants "steal" from a fixed number of available jobs, not realizing that immigrants also drive economic growth and create new jobs. A fervent nationalist might believe that any progress made in another country necessarily translates to their own being worse off, rather than sharing in those benefits. A racist might conclude that any assistance given to other groups necessarily translates to a loss for their own. Not every situation is zero sum, yet this has become a pervasive worldview and an overused metaphor for personal and political relationships, undermining trust and hindering the cooperation that's been so crucial to humankind's success.

Game theory itself is not to blame for the zero-sum bias—von Neumann simply described a specific class of game when he coined the phrase. People harbored some version of this bias long before mathematicians formalized the concept. But because of game theory's privileged position in the halls of academia, people may mistake this folk notion for established truth and use game theory to justify or excuse their distorted views. A person's susceptibility to zero-sum thinking depends on their cultural and personal history. White Americans are more likely than Black Americans to see racial status as zero sum. Men are more likely to see gender as zero sum and oppose fair workplace policies. The countries whose citizens are more prone to zero-sum thinking tend to have lower GDPs, reduced pluralism, diminished government efficacy, and fewer human rights. Individuals prone to zero-sum thinking tend to be lacking in positive emotions and life satisfaction. Zero-sum thinking results in antagonism between individuals and groups, and it can drive people to justify immoral behaviors or make bad decisions based on incorrect assumptions, as it blinkers them to win-win solutions. Given the complexity of human psychology, it's hard to work out the arrow of causality, to determine whether zero-sum thinking causes worse economic outcomes or vice versa. We don't know if the bias is something evolutionarily installed or learned over the course of development. Perhaps it's inherited from overly cynical cultural mythology or learned from playing zero-sum games like chess and misapplied to the real world.

This is not a theoretical problem. It's clear that what people believe makes a big difference in how they behave. In her book *The Sum of Us*, political strategist Heather McGhee details what zero-sum thinking has cost Americans across the racial divide. The notion that nebulous institutions like the market and society are zero sum is simply wrong. Markets are positive sum by nature, and widescale social cooperation has gotten humans where they are today. Yet zero-sum ideological arguments abound in political rhetoric, and some voters bring a zero-sum mindset to the polls.

McGhee details several ways that white Americans have voted against their own self-interest because they mistakenly believe that social or economic gains for Black Americans will result in losses for white Americans. White support for government assistance programs started dropping in the 1960s, once Black Americans stood to be beneficiaries as well. McGhee relates this to psychology research indicating that people are more concerned with their relative position than with their absolute well-being. For example, in a study of game-play, players were rewarded for mutually cooperating. Yet many players still defected for a lower reward, so long as it meant they'd get more points than their opponents. It's possible that the research context influenced these behaviors. People usually play games to win, so players in these laboratory scenarios might have carried over an expectation to win rather than simply collecting meaningless points. These behaviors don't necessarily reflect anything about human nature; rather, they reflect the expectations and values of the given game or culture. If people believe they live in a zero-sum world, they will make choices in line with that belief. The fault is not necessarily inherent in us but in our beliefs about what games we're playing.

It's troubling that, despite its empirical shortcomings, game theory has remained the gold standard of reasoning in certain circles, used to forecast human behaviors a priori. Take ecologist Garrett Hardin's famous 1968 essay, "The Tragedy of the Commons." In it, Hardin uses game theory to invoke Malthusian doom about the effects of population growth. He imagines a grazing commons shared by local herdsmen. As their population grows, they introduce more animals to the land. "As a rational being, each herdsman seeks to maximize his gain," Hardin writes. "The only sensible course for him to pursue is to add another animal to his herd. And another and another. But this is the conclusion reached by each and every rational herdsman sharing a commons." Each herdsman is incentivized to add as many cattle as possible, whereas the cost of overgrazing is shared collectively. The herdsmen collectively ruin the pasture because each individual wants to better their position.

Hardin concludes that "freedom in a commons brings ruin to all" and therefore declares that the "freedom to breed is intolerable." Hardin goes on to argue that reproductive rights should be restricted to prevent overpopulation. He presents this outcome as inevitable: "The inherent logic of the commons remorselessly generates tragedy." The notion that people are helplessly beholden to maximize their personal utility at all costs is a common conceit of those wielding game theory.

Hardin wasn't describing reality, of course, but the abstraction of a maximizing function: the logical consequence of a mathematical tautology. The solutions he imagined were similarly impoverished. Governments have since used his ideas as an excuse to dissolve or privatize commons. His arguments were motivated not by logic but by racism: the Southern Poverty Law Center lists him as a white nationalist in its extremist files. In a 1997 interview, he confessed that "my position is that this idea of a multiethnic society is a disaster." It would be ideal, he writes in "The Tragedy of the Commons," if "the children of improvident parents starved to death," such that "overbreeding brought its own 'punishment' to the germ line." He later called his philosophy "lifeboat ethics," arguing, in effect, that to keep their lifeboat afloat, the global rich should jettison the poor—especially poor immigrants seeking to "steal" the wealth of developed nations.

The rigorously data-driven work of the economist Elinor Ostrom occupies the other end of the spectrum from Hardin's tortured logic. Three years before Hardin published his essay speculating on how an imaginary commons might work, Ostrom published her PhD dissertation detailing how commons work in reality. Ostrom continued to devote her career to studying how people govern common-pool resources, including the management of pastures in Africa, irrigation systems in Spain and Nepal, and fisheries in Maine and Indonesia. She found that locals have created diverse institutions for overseeing these natural resources. Though not all of the institutions she studied were sustainable, she identified common design principles present in most stable management systems. Ostrom went on to become, in 2009, the first woman to win the Nobel Prize in Economics.

These were not rarified thought experiments, floating in a mathematical realm removed from reality. Ostrom collected meticulous data. As Vincent, her husband and longtime collaborator, put it, their method combined "formal approaches, fieldwork, and experiments in order to 'penetrate' social reality rather than to use formal techniques to 'distance' ourselves from it." Humans, Ostrom argues, are "complex, fallible learners who seek to do as well as they can given the constraints that they face and who are able to learn heuristics, norms, rules, and how to craft rules to improve achieved outcomes." People collectively work out globally beneficial outcomes by cleaving to values like reputation, trust, and reciprocity—not by discarding these values. It's important to note that Ostrom's work does not imply that people always sustainably solve collective problems. We wouldn't have destroyed so much of the planet if that were true. However, Ostrom's case studies provide evidence-based hope that collective governance can work and often does better than private or government control. "We are neither trapped in inexorable tragedies nor free of moral responsibility," Ostrom writes.

Ostrom's resistance to a priori models of human behavior was (and continues to be) in direct opposition to the zeitgeist of modern economics research. She claimed that her work was criticized for being "too complex." She countered that we must "be able to understand complexity and not simply reject it." We should regard the diversity of social institutions as a natural resource that deserves protection, much like biological diversity. Still, the human impulse to simplify is hard to overcome: "I still get asked, 'What is *the* way of doing something?' There are many, many ways of doing things that work in different environments." Ostrom's ethos has been facetiously reduced to an oversimple law bearing her name: "A resource arrangement that works in practice can work in theory."

Despite its flaws, the conceptual tragedy of the commons has persisted in cultural memory, reduced to a popular catchphrase. It's a disturbing example of how a supposedly objective thought experiment can be misappropriated and used to mask politically motivated arguments. Hardin's ideas have been used to justify privatization as the solution to natural resource management, fetishizing theoretical efficiency over important goals like sustainability. His ideology has its roots in twentieth-century eugenics—the same horror that von Neumann hoped game theory would countervail. Hardin's rhetoric has inspired racist, nativist, and anti-welfare sentiments and policies for decades. One current incarnation of Hardin's argument is known as the "great replacement" theory, which suggests that white populations are being supplanted through immigration and low birth rates. This notion has inspired several domestic terrorism events, including the 2017 Charlottesville riots and the 2022 mass shooting in Buffalo. Hardin himself aggressively lobbied Congress for anti-immigration policies and cuts to social services. Neo-Malthusian arguments, advanced in Hardin's work and Paul Ehrlich's *The Population Bomb*, went on to have a chilling effect on global politics. China adopted its one-child policy, resulting in millions of forced abortions. The World Bank and the UN pressured countries including India, Egypt, Tunisia, Peru, and Bangladesh to incentivize—often coercively—mass sterilizations. In India alone, tens of millions of Indian men and women have been sterilized to date. Thousands have died after hastily performed procedures in substandard medical facilities, most of which are performed on poor women from lower castes.

What we believe—about ourselves, about other players in the game, about the structure of the game worlds we inhabit—influences our preferences and choices. Our models of the world, like Hardin's tragedy, encompass the solutions we envision for it. Because the world is interrelated, our choices impact the choices and lives of other people.

They transform the terms of the game for other players—the same way the prisoner's dilemma can become a lover's delight. We may become victims of game theory, making worse choices because we assume that other players are rational strategists. Or we might find ourselves battling to buy a dollar for two dollars, as with Shubik's dollar auction.

Game theory has little to teach us about human behavior, except, perhaps, how desperate people are to understand one another. The fact that it's been so wrong might explain how psychologists have faithfully mined it for decades. Dopamine neurons are oracular, translating the brain's internal model of the world into predictions about what will happen next. These neurons register prediction errors— deviations from their predictions. Large deviations signal to the brain that there is something to learn, that its model is somehow off. It's an elegant system, but it's also vulnerable to manipulation depending on the beliefs we adopt. People are famously susceptible to shallow ideas and world models: the facile TED Talk, the one-sentence airport-book thesis. Unlike animals, humans can share their internal models using language, and others can adopt, adapt, or reject them for their own use. The simplicity of an idea is something like a virus's infectivity. Oversimple ideas infect many people precisely because they are so simple. The easier an idea is to communicate, the further it spreads. If the idea is somehow surprising or counterintuitive, so much the better— the brain registers this as especially salient. This is why authors of airport books and edgelord bloggers are rewarded for courting surprise with simplistic arguments. If an idea is both simple and surprising, it is doubly virulent. The fitness of an idea or meme, measured by its reach, has nothing to do with its correctness and everything to do with how easily it can be grasped, remembered, and transmitted.

The most unfortunate part is that if we take one of these wrong ideas as truth—perhaps because it is offered by someone with the right social, religious, or academic credentials—it can become almost unfalsifiable. If we incorporate a bad idea into our world model, it may generate lots of prediction errors, because it's a poor model of the world. Yet if we value the idea enough, we will take these prediction errors not as evidence against it but as evidence that we are learning from it. For argument's sake, let's take the game-theoretic model of human rationality. The model posits that humans are rational: they act like self-interested value maximizers. When this model was tested empirically, researchers discovered that humans deviate from it in many significant ways. Instead of discrediting game theory's account of human behavior, this was taken to indicate that humans aren't rational. Many behavioral economists have made their careers by documenting the many ways that humans deviate from rational assumptions. It may seem like we're learning something from game theory because it has generated so many prediction errors. Perhaps it's valued precisely because it generates so many surprising differences between real humans and game theoretic agents, and surprising results play a significant role in getting papers published. But couldn't this also just mean that it's a poor model of human behavior? Why do we insist on calling the ways humans differ from a mathematical ideal biases instead of taking them as evidence that game theory is not a good model of people? Game theorist Robert Sugden writes:

> It is as if decision-makers are held to be at fault for failing to behave as the received theory predicts, rather than that theory being faulted for failing to make correct predictions. It seems to me that, in their desire to make normative pronouncements, many of my fellow behavioral economists are not taking the psychological evidence sufficiently seriously.

One of the concerns shared by philosophers, naturalists, sociologists, poets, soldiers, and civilians alike is humankind's "true" nature.

Perhaps the historical lack of consensus should itself be a lesson: our nature has been hard to pin down because it isn't fixed. Our hallmark quality is plasticity. We are, to borrow a phrase from historian Johan Huizinga, *Homo ludens*, man the player. We slip in and out of identities, value systems, and agencies the way players inhabit the strictures of a game.

Games have always been about discovering who we are. Friends and family members agree to suspend everyday reality to temporarily compete. Games sharpen our ability to understand the goals and intentions of the people we interact with and yield new insights into otherwise hidden aspects of their personalities. They free up our constrictive senses of self, loosen the straitjacket of identity, and allow us to explore virtual worlds and act in ways we might not otherwise. Think of the unassuming aunt who's a merciless Monopoly shark, or the vegan friend with a massive virtual body count. Games allow us to learn new strategies and ways of thinking from one another without necessarily speaking the same language. Perhaps they evolved for just this reason. In the same way that astronomy explains the motion of stars and planets, games are a science of selves: contained experiments that teach players the principles of interaction. Because we have documented the position of Saturn for thousands of years, we know where to point our telescope tomorrow night. Because we have honed our strategies in play, we can both meet the challenges of a competitor and better anticipate the needs of an ally.

Games don't just show us who we are—they reward certain behaviors. People will be only as good as the games they're incentivized to play. They'll be as good as they expect the games themselves—and their fellow players—to be. Humans are not a fixed collection of preferences. They learn, and their behaviors are shaped in response to the games they find themselves in. Game theory is a pristine branch of mathematics and a brilliant model underpinning the dynamics of complex computer networks, evolutionary dynamics, and certain machine-learning algorithms. It tells us nothing about human nature. Game-theoretic agents have no sense of *us*, no concern for community. People do.

Game theory's power lies in its expressiveness: it can convey dynamic stories with simple mechanics. Unfortunately, game theory has been used to launder dubious beliefs and present racist, nativist policies as objective or inevitable. Some of these sinister fables have been taken out of context and integrated into the foundations of accepted common sense. If game theory continues to be used as a model of human behavior, it needs major revisions to reflect the fact that most people aren't singularly focused on personal gain. Even then, we'd be better off modeling people for what they are: flexible learners. People are more than their neural circuits, of course. But we'll have better luck understanding ourselves in light of our biology than by way of mathematical abstraction. In fact, we may be worse off if we apply lessons taken from these mathematical models to our real-world decisions. People make choices influenced by contextual clues, whether they believe themselves to be playing the Wall Street Game or the Community Game. So what games are we truly playing? And can we design better ones?

A Map That Warps the Territory

The chief reason warfare is still with us is neither a secret death wish of the human species, nor an irrepressible instinct of aggression, nor, finally and more plausibly, the serious economic and social dangers inherent in disarmament, but the simple fact that no substitute for this final arbiter in international affairs has yet appeared on the political scene.

HANNAH ARENDT

With the advent of nuclear weapons in World War II, nations scrambled to find a new diplomatic equilibrium. This was hardly the first time the balance of international power had been disrupted by science. Every new military technology radiates a certain inevitability, as though it exists latent in some Platonic realm, patiently awaiting discovery—and rediscovery. Long before napalm burned its first victims alive in 1943, the Byzantines invented an incendiary substance known as Greek fire. First documented in 672 CE, it protected Constantinople from outsiders for seven hundred years until its recipe was lost. The seemingly supernatural technology was unstoppable; even water couldn't quench it. Around the same time, Chinese and European engineers independently perfected the cross-bow, which was eventually capable of piercing armor. Medieval knights petitioned to outlaw the deadly weapon, and in 1139, Pope Innocent II forbade its use against Christians—though Europeans still used the technology against non-Christians in the Crusades.

Next, gunpowder would horrify the world. Taoist scholars created the deathly mixture by accident—it was evidently a by-product of their search for the elixir of life. By the eleventh century, the Chinese military had employed gunpowder in a weapon that was something like a flamethrower, and then in a cannon. European and Middle Eastern engineers further developed cannons, then guns, beginning around the thirteenth century. Worse was to come: in the nineteenth century, Alfred Nobel spent his early career perfecting the formula for the explosive substance nitroglycerin, initially intended for commercial mining and construction. He was an avowed pacifist but continued to develop explosives as a military technology because he believed they'd have a chilling effect on warmongering. He assumed their use was too morally terrible to contemplate. In 1891, he boasted to his friend Bertha von Suttner, the famed pacifist author and activist: "Perhaps my factories will put an end to war sooner than your congresses. On the day that two army corps can mutually annihilate each other in a second, all civilized nations will surely recoil with horror and disband their troops."

The idea of a technology that could end war remained a surprisingly persistent conceit. English author Wilkie Collins reflected on it in 1870: "I begin to believe in only one civilizing influence—the discovery, one of these days, of a destructive agent so terrible that War shall mean annihilation, and men's fears shall force them to keep the peace." Military scholar Jan Bloch argued in 1898 that advanced military technologies had rendered war "impossible alike from a military, economic, and political point of view." When aircraft were invented, people speculated that planes might be used for aerial scouting, but the public conscience would never permit them to drop explosives on cities. Within a few decades, "the public conscience became unaccountably numbed," wrote nuclear strategist Lewis Strauss. The horrors of World War I made another global conflict seem inconceivable. Yet as Hitler rose to power, journalists nervously speculated on the return of the "oldest and most blood-drenched fallacy in History: 'the war to end wars.'"

Von Neumann eagerly volunteered his intellect to the Allied effort. In 1943, he worked with the British military to apply minimax techniques to determine the best places to sweep for German undersea mines. After witnessing an accounting machine at work during his visit, von Neumann confessed that he'd developed an "obscene interest in computation techniques." He went on to invent several major advances in computing, and the prevailing modern computer architecture is based on his design. Operations research, established around this time, was founded on minimax and other computer optimization techniques. This new field, scientists promised, would streamline complex systems, further underscoring game theory's practical promise. The public fretted about the new field of computing, however. Rumors circulated that almost half of the scientists working on these mechanical brains had gone mad, their minds obediently compounding dizzying arithmetic calculations as though caught in an indeterminate loop.

The same year, von Neumann was recruited to join the Manhattan Project at Los Alamos to explore the physics of detonation. German scientists, intelligence had it, were already working on an atomic bomb; America would have to invent it first. Von Neumann brought his new computational prowess to bear on modeling the dynamics of the bomb. The military trusted him with weighty political decisions as well. He served on the committee that recommended target cities, balancing their military utility with the psychological impact of their destruction. Von Neumann was motivated by his desire to preserve political systems that protected intellectual freedom, which he saw as the main route to improving the human condition. "I'm convinced, in fact," wrote his daughter, Marina, "that all his involvements with the halls of power were driven by his sense of the fragility of that freedom." In 1944, German forces occupied Hungary. Rudolf Ortvay, von Neumann's lifelong friend and faithful correspondent, took his own life in Budapest a year later.

The nuclear destruction of Hiroshima and Nagasaki ended World War II and imperiled the globe with a terrible new peace. The Americans had developed a world-ending technology to which diplomacy, ethics, and political strategy had yet to catch up. Missiles, aircraft, and nuclear weapons had vastly accelerated the dynamics of war. Decisions that could formerly be mulled over for months, like the speed of ground troops, now had to be made in days, even hours. Von Neumann wrote: "It will not be sufficient to know that the enemy has only fifty possible tricks and that we can counter every one of them, but we must be able to counter them almost at the very instant they occur." Thinkers scrambled to invent a totally new science, forged out of existential necessity: a calculus of war. Decisions would have to be made with lightning speed and superhuman accuracy, unfettered by human bias or moral scruples. Many military strategists would come to adopt game theory, the "science of conflict," as a cognitive prosthesis.

The mathematician had grown up playing Kriegsspiel, and game theory was, in many ways, his answer to German technology. Kriegsspiel offered a bird's-eye view of a battlefield, while game theory provided a bird's-eye view of conflict itself. Game theory, von Neumann promised, would clarify the rational decisions with which humans could no longer be trusted. Defensive reflex would be replaced with calculation. War, claimed game theorist Anatol Rapoport, would be elevated to the level of games, "where the intellect has a chance to operate."

President Truman's administration considered the nuclear bomb a weapon of terror, a technology of last resort distinct from the conventional arsenal. Yet many of his military advisers, including von Neumann, argued it should be wielded liberally. Von Neumann described himself as an unusually hawkish person, and "violently anti-communist." He argued for the then-popular idea of a "preventative war," intending to undermine the Soviet Union by launching a nuclear attack before they could develop their own nuclear weapons. "For progress," von Neumann wrote, "there is no cure." This thinking was in part driven by game-theoretic strategic reasoning. If one country develops a weapon, every other country is incentivized to develop the same. "With the Russians, it is not a question of whether but of when," von Neumann argued. "If you say, Why not bomb them tomorrow?, I say, Why not today? If you say, Today at five o'clock, I say, Why not one o'clock?" But it was also motivated by a regret that haunted many Europeans. In 1935, Hitler had declared that Germany would violate the Treaty of Versailles and rearm itself. Many believed that if European powers had not appeased the leader and instead had preemptively attacked the still-weak country, the Holocaust and World War II might have been avoided entirely. Perhaps the same would be true of potential conflicts with the Soviet Union.

Von Neumann was unable to convince US military officials to preemptively strike Russia before it was too late. In part through the subterfuge of a fellow Los Alamos physicist, Klaus Fuchs, the USSR soon had a successful nuclear program. On August 29, 1949, the USSR tested their first nuclear bomb in Kazakhstan and quickly spun up an arsenal. The architects of the new world order needed a stable equilibrium point now that disarmament was out of the question. The most famous strategy to emerge from game theoretic arguments was later given the backronym MAD: mutually assured destruction. According to this doctrine, the use of atomic force against another nuclear power should elicit a backlash so severe that it would spell utter destruction for both the attacker and the defender. This all-or-nothing approach would stabilize international relations and deter aggression. MAD is a Nash equilibrium: participants locked in mutual deterrence wouldn't prefer any other strategy. If they initiated conflict, they'd ensure their own destruction. If they disarmed, they'd be vulnerable to annihilation by their enemy.

MAD was not formalized as a doctrine until the 1960s, though it was becoming more practically feasible. After the USSR launched a successful long-range missile test in 1953, von Neumann led a US Air Force committee whose findings underscored the importance of developing long-range missiles that could initiate nuclear strikes anywhere in the world—a technology, the committee warned, the US would have to catch up with the Soviets to develop. The US likewise accelerated its efforts to develop the hydrogen bomb, with even greater destructive power than the nuclear bomb. Citizens worldwide, on whose behalf war was being threatened, were terrified at being drawn into political games they never consented to play, held hostage by their own militaries. They wrote scientists and officials frantic letters, worried that these bombs would blow a hole in the bottom of the sea, set the atmosphere on fire, push up new mountain ranges, cause tsunamis, or destroy gravity. Historian S. M. Amadae points out that game theoretic models of war broke with classical liberal mutual regard—wherein individuals protect their own rights by respecting one another's rights—in favor of pure self-regard.

Following several months of shoulder pain, von Neumann was diagnosed with cancer in 1955. He was soon confined to a wheelchair, then bedridden. He threw himself into his final work: an unfinished manuscript drawing analogies between computers and the brain in which he argued that intelligence is essentially statistical.

Von Neumann spent the last year of his life at Walter Reed, racked with pain and obsessed with his intellectual legacy. He'd played a significant role in inventing the modern computer, early AI, the atom bomb, and weather forecasting. He'd predicted global warming and climate change, axiomatized quantum mechanics, and theorized the existence of machine automata that could replicate and evolve. His work had changed the very trajectory of history. Yet according to his colleague, mathematician Raoul Bott, von Neumann confessed that "he never really felt that he had lived up to what had been expected of him."

This sense of inadequacy haunted him in his final days. He'd grown obsessed with immortality: not only of his intellectual contributions but of his soul. He converted to Roman Catholicism in the last year of his life, to the surprise of all who knew him (he had always been, according to a colleague, "uninhibited by ethical considerations"). Marina, von Neumann's only child, visited him often in the hospital. She surmised that, in his religious turn, he was affirming Pascal's wager and covering his bases in the event that an afterlife exists. He met regularly with the theological scholar and Benedictine monk Anselm Strittmatter, putting his childhood talent with ancient languages to work by reciting penitential psalms in their original Latin.

Religion brought von Neumann little peace. His obituary in *Life* magazine reported that his body, "which he had never given much thought to, went on serving him much longer than did his mind." Rendered immobile, he was forced to witness the dissolution of his intellect through the fog of painkillers and illness. Hoping to chart the progression of his fading mind, he asked Marina to test him on basic addition. As a child, he'd charmed adults by multiplying eight-digit numbers in his head. Now he could hardly retain a short sum. His colleague Edward Teller recounted that "von Neumann suffered more when his mind would no longer function than I have ever seen any human being suffer." He underwent a complete mental breakdown.

Drowning in existential panic, he screamed in hopeless terror every night. The US military stationed a twenty-four-hour security duty outside his hospital room to ensure that he wasn't leaking nuclear secrets in his delirium. His experience was so severe that his wife and brother both wrote living wills prohibiting the use of medical means to keep them alive. Von Neumann died in 1957 at the age of fifty-three. Six years later, Klára Dán walked into the Pacific Ocean and drowned herself.

At von Neumann's funeral, Strittmatter affirmed that the mathematician had never expressed any regret over his involvement with the bomb. The project's director, Robert Oppenheimer, famously lamented his role with a quote from the Bhagavad Gita: "Now I am become Death, the destroyer of worlds." Von Neumann retorted that "some people confess guilt to claim credit for the sin." Yet, according to the monk, von Neumann had grown aware that scientific insight couldn't obviate the need for human values:

He felt with steadily increasing intensity the moral problems bound up with the greatest of modern scientific triumphs. This realization that man was proceeding along a track which might well lead to power above and beyond control was the beginning of an insight into another realm, into a world of values which cannot be ignored, for it is as real and vast, yea more vast than the immense world of the atom he had helped to explore. Here were responsibilities clamoring for attention. He could not avoid the manifold implications of the problem.

Nevertheless, scholars and strategists continued to use game theory as the standard framework for modeling conflict, hoping it would help them navigate the complex labyrinth of international relationships in the era of nuclear diplomacy. So continued the dream that we might move from fighting wars to computing them. In his 1960 book, *The Strategy of Conflict*, the political theorist and RAND consultant Thomas Schelling explores the idea of modeling the nuclear standoff between the US and USSR as a prisoner's dilemma. How can we compel players caught in a prisoner's dilemma to cooperate? Schelling argues that this requires an ironclad mutual commitment, secured by punishment should either party defect. The threat of mutually assured destruction, he contends, serves this purpose. Should one nation defect, the other promises to swiftly punish them. No actual conflict ever need arise; instead, the "vicious diplomacy" of intimidation will stabilize relations in a kind of continuous bargaining. Counterintuitively, aggressive posturing could secure peace. Perhaps, as Lasker had hoped, game theory would render war superfluous, substituting the mere threat of nuclear force for kinetic war. Schelling was awarded the 2005 Nobel Prize in Economics for his work modeling conflict as a form of bargaining.

Not everyone agreed with Schelling's assessment. In 1960, Herman Kahn, a fellow RAND researcher, published *On Thermonuclear War*, a rambling treatise detailing how the bomb had changed the nature of conflict. He intentionally courted horror in his readers. MAD, Kahn reasoned, is uniquely vulnerable to rogue personnel, accidents, and sabotage. His book inspired Stanley Kubrick's 1964 movie *Dr. Strangelove*, whose eponymous doctor—associated with the satirical BLAND Corporation—is a pastiche of several strategists, including von Neumann and Kahn. In the movie, US forces learn that the Soviets have built a nuclear Doomsday Machine. Its detonation will be automatically triggered in response to any nuclear attack, leading to radiation levels so high that the earth will be rendered uninhabitable. Unfortunately, a rogue US general attacks the Soviets before they have announced this deterrent mechanism, thus ending the world. The movie illustrates MAD's vulnerability and—more fatally to military strategists obsessed with rationality—a flaw in its sequential logic. Kahn demonstrated:

> Let us assume that the president of the United States has just been informed that a multimegaton bomb has been dropped on New York City. What do you think that he would do?
>
> When this was first asked in the mid-1950s, the usual answer was "Press every button for launching nuclear forces and go home." [. . .]
>
> The dialogue between the audience and myself continued more or less as follows [. . .]:
>
> **Kahn:** What happens next?
> **Audience:** The Soviets do the same!
> **Kahn:** And then what happens?
> **Audience:** Nothing. Both sides have been destroyed.
> **Kahn:** Why, then, did the American president do this?

It's not rational to act on the threat of massive retaliation. Is ending the world a sensible punishment? But MAD requires both sides to credibly and totally commit to utter destruction. The enemy must be assured that their opponent can counterattack and will absolutely follow through. Only an amoral agent could carry out such a retaliation. To maintain this threat, each side would need either a precommitted computer system or a "madman" in charge of their arsenal.

Having recognized this, Soviets and Americans alike worked on committed computer systems that, like Kubrick's Doomsday Machine, could automatically and irrevocably retaliate if provoked. The idea featured heavily in popular media. The 1951 sci-fi classic *The Day the Earth Stood Still* imagined an alien coalition that granted absolute power to their robotic police, who thoroughly extinguished every hint of aggression. The technology of the time was nowhere near capable of automated decision-making, but politicians adopted the fantasy as reality. Schelling recalled Soviet premier Nikita Khrushchev threatening an American diplomat: "Your generals talk of maintaining your position in Berlin with force. That is a bluff. If you send in tanks [. . .] our rockets will fly automatically." This was meant to underscore the credibility of their threat. The Soviets wouldn't be able to stop their missiles even if they wanted to.

Luckily, such a system never existed. We avoided outright nuclear apocalypse in several near misses during the Cold War thanks to human, not machine, reasoning. In 1962, Khrushchev secretly installed nuclear missiles in Cuba to deter the US from invading the tiny nation. His aim was not to court war but to maintain, as he put it, a "balance of fear" to prevent one. Yet when US military officials discovered his act, the ensuing political volatility nearly sparked a nuclear showdown. Thanks to clear-eyed diplomacy—not game theory—both powers eventually backed down and agreed to dismantle a suite of atomic weapons pointed at one another. In 1983, the Soviet officer Stanislav Petrov narrowly avoided a full-scale nuclear attack when he correctly interpreted an incoming missile warning as a false alarm. Procedure dictated that he should alert his superiors about the alarm so they could initiate nuclear protocols. He didn't raise the alert, reasoning that the missile warning was more likely a technical error than a genuine attack. History proved Petrov right—a high-altitude solar reflection had tripped the alarm. In retrospect, it's clear what a terrible mistake it would have been to abdicate this kind of responsibility to a machine protocol. Petrov's act of heroism underscores Kahn's criticism of MAD. Any sane official would refuse to carry out orders to destroy the world, and would be rightly celebrated as a savior.

In lieu of automated systems, American strategists argued, our leaders could simply eschew sanity. A nation could make a credible threat of retaliation by putting an amoral leader in charge of the nuclear trigger. Counterintuitively, a strategy developed along the ley lines of pure rationality required irrationality to be credible. Schelling had likened this diplomacy to a game of chicken. Crucial to winning chicken is displaying an attitude of complete depravity. Once one's reputation is lost, it can't be regained. He described brinkmanship as "the tactic of deliberately letting the situation get somewhat out of hand, just because its being out of hand may be intolerable to the other party and force his accommodation." Imagine, for instance, that two prisoners are chained together at the ankle and placed on the edge of a cliff, tasked to battle for their freedom. Only one can be released. It wouldn't be credible if one of these prisoners threatened to push the other, as that would result in both falling. Instead, one could leverage unpredictability: dancing at the edge, inching closer and closer to the precipice until their chained partner capitulates. Irrational-seeming bluster is, in this case, rational. In this vein, America's war in Vietnam was intended to demonstrate the credibility of America's threat to the rising communist tide. Accordingly, President Richard Nixon intentionally cultivated his reputation as trigger happy: "I call it the Madman Theory," he explained. "I want the North Vietnamese to believe I've reached the point where I might do anything to stop the war."

Nixon and his advisers had rediscovered an age-old strategy wielded by monarchs for millennia, from the Great Sun presiding over the Natchez of the southeastern US to the absolutist Sun King, Louis XIV. By committing some moral atrocity, leaders signaled that they were not bound by traditional morality, though they would presume to impose moral order on their kingdoms. Nixon's eventual successor, the devout Jimmy Carter, didn't possess Nixon's sinister credibility. It was clear that his Christian faith would compromise any act of retribution. Carter's administration therefore shifted from the MAD doctrine to NUTS: nuclear utilization target selection.

NUTS, advocated by Kahn, is based on the idea that nuclear force needn't end the world. Nations can, in theory, use atomic weapons in the same escalatory manner as conventional weapons. It's this strategy of limited nuclear war that Putin has threatened to use against Ukraine. In a NUTS scenario, officials employing tactical nuclear weapons should target military bases and infrastructure before escalating to cities or incurring civilian casualties. But it was immediately clear to Carter's secretary of defense, Harold Brown, that any use of a nuclear weapon would "very likely" escalate to general nuclear war. Military advisers Spurgeon Keeny Jr. and Wolfgang Panofsky concluded that "the unprecedented risks of nuclear conflict are largely independent of doctrine or its application." No matter whether the US officially subscribed to MAD or NUTS, any use of nuclear force would escalate to total destruction.

As discussed in the previous chapter, players will behave differently depending on the game they believe they're playing. Because strategists argued that international diplomacy resembles a prisoner's dilemma, they advocated for increasingly adversarial negotiation tactics. Worse, players should intentionally act irrationally, or relinquish control to remorseless automatons, in order to guarantee global destruction more credibly. Only a few top-ranking advisers fully grasped the precarity of atomic diplomacy. Herbert York was a nuclear physicist of Mohawk descent, the first president of the Lawrence Livermore National Laboratory, and a key player in developing the US nuclear program. He warned against the US government's attempt to solve sociopolitical problems with technological solutions, contending: "There is no such thing as a good nuclear weapons system. There is no way to achieve, in the sound sense, national security through nuclear weapons." Technological exuberance only threatened global safety, locking both sides in an escalating bid to increase their military power. We were all less secure as a result. York worried that doomsday would be ushered in by either a computer program or "a preprogrammed president who, whether he knows it or not, will be carrying out orders written years before by some operations analyst."

York was not alone in his concern. When President Dwight Eisenhower left office in 1961, he warned the public about the growth of the military-industrial complex, predicting that "public policy could itself become captive of a scientific technological elite." However, he maintained that the military-industrial complex and the technological elite were ultimately necessary. His successor, President John F. Kennedy, embraced a technological approach to war when he appointed the management consultant Robert McNamara—a RAND acolyte— as his secretary of defense. McNamara had served in the Office of Statistical Control during World War II, analyzing bomber efficiency. He and colleagues from that office formed a consulting group after the war. They brought their analytical acumen to the Ford Motor Company, utilizing optimization principles to reinvigorate the floundering corporation. Having risen through the ranks of Ford to ultimately serve as CEO, McNamara accepted the position in Kennedy's cabinet, though he confessed concern that he lacked sufficient military expertise. Kennedy assuaged McNamara's fears: "We can learn our jobs together. I don't know how to be president either."

McNamara brought his gospel of technological optimization to bear in the Vietnam War, applying rational management principles to the disastrous campaign. His team of civilian consultants boiled military strategies down to their quantifiable bones: What weapons, where, and how many? Decisions were made according to predicted performance metrics and budgetary considerations: number of lives lost, percentage of infrastructure damaged. He famously eschewed military experts' advice in favor of his analysts' synthetic rationality. RAND strategist Bernard Brodie recounted that McNamara was "plainly in love" with his method of systems analysis: "He gloried in the graphs, multicolored, in layer upon layer, and rejected the 'poetry' of those who sought to introduce a little political intuition." However, the seemingly scientific technique led to a vast increase of US forces in a war with no clear win state. The tools he'd used to rescue Ford failed in military applications.

McNamara appointed Charles Hitch, the head of economics at RAND, to serve as assistant secretary of defense. Hitch was more familiar with game theory's empirical failings, which significantly dimmed his confidence in modeling people as rational agents. As such, he was more cautious about the wanton use of slick analytics. At RAND, he'd already concluded that game theory was "disappointing" in military applications. He advocated for more holistic analyses:

The future is uncertain. Nature is unpredictable, and the enemies and allies are even more so. [The analyst] has no good general-purpose technique, neither maximizing expected somethings, nor *max-min*ing, nor gaming it, to reveal the preferred strategy. How can he find the optimal course of action to recommend to his decision-maker? The simple answer is that he probably cannot.

A war of theoretical efficacy and efficiency was no match for guerilla tactics in Vietnam, partly because the Americans were optimizing metrics that didn't apply to the developing nation. American bombs couldn't devastate military or industrial infrastructure because Vietnam had little of either. The US strategists had committed the cardinal sin of projection: assuming that their opponent's utility function was the same as their own. By 1965, it had become clear to McNamara that his optimistic computer projections were wildly out of line with conditions on the ground. The Germans had used Kriegsspiel simulations to make accurate predictions in their military campaigns, but America's gamed-out simulations were failing because they got the rules all wrong. Vietnamese guerilla fighters were changing the very nature of warfare. In a retrospective letter to *Foreign Policy*, Brodie admitted that he and his fellow analysts "fell flat on their faces over the past decade in trying to predict the character and outcome of the Vietnam War. [. . .] It is an important and disturbing conclusion."

McNamara was showing signs of stress. President Lyndon B. Johnson—Kennedy's successor—accused him of "cracking up." McNamara left, or was forced to leave, his position as secretary of defense in 1968. On his last day as defense secretary, he tearfully begged Johnson to stop sending troops to Vietnam: "What then? This goddamned bombing campaign, it's been worth nothing, it's done nothing, they've dropped more bombs than in all of Europe in all of World War II and it hasn't done a fucking thing." Nothing changed. In a 1969 attempt to aid the growing anti-war effort, RAND game theorist Daniel Ellsberg released seven thousand pages of classified documents that he'd gained access to through his military clearance. Now known as the Pentagon Papers, they revealed a program of lies perpetrated by several presidents and their military attendants, and a series of war crimes committed against civilians across Southeast Asia. A majority of American citizens already disapproved of the war. One military official noted that "a feeling is widely and strongly held" around the country that "'the Establishment' is out of its mind." Yet the Vietnam War continued for another six years. American military officials considered the war central to maintaining the credibility of their deterrent threat against the USSR.

Understandably, game theory was not well loved by the American public. Many saw it as a tool to justify nuclear war, its "objective" rationality endorsed by a shadowy think tank funded by a war supplier seeking to sustain market demand after the end of World War II.

Nations became deadlocked in the logic of strategic rationality, re-warded for abandoning traditional ethics to render their threats more trustworthy. Oppenheimer was not shy with his criticisms. "What are we to make of a civilization," he asked, "which has always regarded ethics an essential part of human life, and [. . .] which has not been able to talk about the prospect of killing almost everybody, except in prudential and game-theoretic terms?"

The threat of nuclear annihilation looms to this day. Kubrick got it right in *Dr. Strangelove*: a single rogue official with nuclear clearance could end the world. In 2020, Russia amended its constitution to give President Vladimir Putin primary control over its nuclear arsenal. The same power has always been available to American presidents, who hold unilateral authority to initiate a nuclear attack of global propor-tions. In 1974, Nixon bragged to reporters: "I can go into my office and pick up the telephone and in twenty-five minutes seventy million peo-ple will be dead." Citizens and politicians alike have petitioned to end this precarious privilege. It seems reasonable to give at least one other official veto power over the US president, given the ease with which a single person might make an apocalyptic decision based on a momen-tary lapse in judgment. Various presidents' personal weaknesses am-plified this danger. Kennedy, for instance, was addicted to painkillers; Nixon, to alcohol. Trump posted mercurial tweets threatening nuclear aggression against North Korea. Any American president could, in theory, precipitate apocalypse after a bender or to satisfy a personal grievance.

Whereas game theory inflamed international tensions during the Cold War, a different game would come to ease them. In the early 1980s, nuclear war fears were building up after the relative calm of the 1970s. At the invitation of the US secretary of defense, Schelling organized an extended war-game simulation known as *Proud Prophet*. Over the course of several weeks, two hundred top-ranking military officials and politicians worked through various scenarios of limited and unlimited nuclear war in Asia, Europe, and the Middle East. Every outcome was relentlessly grim. The least deadly of the bunch resulted in five hundred million hypothetical deaths and more to come, given that nuclear fallout would render the northern hemisphere uninhabitable. Escalatory, "limited" nuclear war—the same tactic that military strategists still discuss in relation to the conflict in Ukraine—mounts inexorably toward mutual destruction, wiping all life off earth. Everyone involved in *Proud Prophet* was deeply disturbed. It was clear how inadequate America's existing strategies were, and how unprepared its leaders.

Proud Prophet's harrowing outcomes were compounded by Reagan's reaction to the recently released movie *WarGames*, which he'd screened at Camp David. In the film, Matthew Broderick plays David Lightman, a computer-savvy teenager who, thinking he's hacked a game company's computer, accidentally hacks into the military's nuclear simulation engine and nearly triggers a global war. The system he gains access to, WOPR (War Operation Plan Response), constantly churns through potential battle scenarios to "learn from mistakes we couldn't afford to make." Thinking it a video game, Lightman unwittingly invites WOPR to initiate World War III. WOPR obediently plans and executes an attack against the USSR. Once the attack mounts, Lightman finds that he's unable to stop the computer from carrying it out. Instead, he invites WOPR to play tic-tac-toe. After a series of draws, WOPR realizes that sometimes "the only winning move is not to play" and disarms its missiles at the last minute. *WarGames* spooked Reagan. He revamped national security measures and passed anti-hacking legislation that still exists, in some form, today. In light of *Proud Prophet*'s outcomes, Reagan's administration changed America's nuclear rhetoric, focusing on defensive, de-escalatory measures and prioritizing talks to freeze nuclear proliferation. Together, *Proud Prophet* and *WarGames* were a surprisingly effective deterrent to military escalation. The threat of atomic diplomacy eased.

The general public continued to distrust game theory. Kahn, the strategist responsible for the limited nuclear strategy NUTS, complained of the criticism he received for his books. This criticism was not, he claimed, concerned with the merit of his arguments themselves. Instead, it focused on whether it was immoral to write, or even think, about fighting a thermonuclear war. He replied to the critics railing against the "icy rationality" of game theory. "Would you prefer a warm, human error?" he asked. "Do you feel better with a nice emotional mistake? We cannot expect good discussion of security problems if we are going to label every attempt at detachment as callous, every attempt at objectivity as immoral."

But lost in Kahn's defense of game theory's objectivity is that it's tautological. Facts are not the same thing as mathematical truisms. Game theory cannot perfect our decisions or safeguard against poor choices. It does not inoculate the world from the need for human values. Even capitalism's patron saint, Adam Smith, didn't advocate for unfettered selfishness. He argued that individual self-regard can be collectively beneficial—but he also took morality for granted. Customers can still trust selfish merchants because people generally want to uphold their reputations as honest and moral. "How selfish soever man may be supposed," Smith writes, "there are evidently some principles in his nature, which interest him in the fortune of others, and render their happiness necessary to him, though he derives nothing from it except the pleasure of seeing it."

In his book *Economic Fables*, mathematician Ariel Rubinstein likens game theory—mathematical abstractions of complex situations—to fables with oversimple morals. We should be careful not to mistake these for reality. Game theory should be studied for its mathematical beauty, but it cannot pretend to tell us about the real world. Rubinstein argues that even many academics don't fully understand it and often fail to communicate its limitations. This would not be an issue if game theory remained a purely academic pursuit. But it's been adopted for use in modern systems that impact billions of people worldwide. Rubinstein writes:

> I think it's a very tempting idea for people, that they can take some-thing simple and apply it to situations that are very complicated, like the economic crisis or nuclear deterrence. But this is an illu-sion. Now my views, I have to say, are extreme compared to many of my colleagues. I believe that game theory is very interesting. I've spent a lot of my life thinking about it, but I don't respect the claims that it has direct applications. [. . .] I have not seen, in all my life, a single example where a game theorist could give advice, based on the theory, which was more useful than that of the layman.

Game theory was, in a way, a precursor to AI: a synthetic rational-ity that promised to augment, if not entirely replace, human decision-making. In a 1962 essay, the mathematician Anatol Rapoport warns of game theory's potential for abuse. He criticizes how strategists model wars as zero-sum games ("they are not!"). Game theory can only pre-scribe the best strategies for specific classes of games. Moreover, it's best at outlining the limits of what can and can't be done, but it can rarely prescribe practical actions. "It teaches us what we must be able to do in order to bring the intellect to bear on a science of human con-flict," he writes. We can hope to move combat from the level of reac-tive reflex to the level of games. But game theory's idealized solutions leave out essential considerations like honesty, responsibility, and sim-ilar virtues. Without these "extra-game-theory" considerations, con-flicts like the prisoner's dilemma lead to an impasse. Getting out of such deadlocks requires traditional interpersonal skills like communication, listening, and perceptivity.

The hope that synthetic reasoning will replace human decision-making—and lessen the burden of our moral responsibility—persists in today's war automatons. Widely hailed as one of the greatest early video games, Silas Warner's 1981 *RobotWar* is a game with no players. It heralded modern drone combat, imagining a universe where human lives are replaced, in the accounting of war, with machines locked in a struggle for survival. "WELCOME TO THE BATTLEFIELD OF THE FUTURE!" its manual begins.

It is the year 2002. Wars still rage, but finally, they have been officially declared hazardous to human health. Now, the only warriors are robots—built in secret and programmed to fight each other to the death!

Your country has just developed the most efficient battle robot to date. It should be unbeatable—but part of its micro-computer "brain" is still blank. Only when a strategy is programmed into its memory will the robot be able to fight.

The task set before you is:

TO PROGRAM A ROBOT,

THAT NO OTHER ROBOT CAN DESTROY!

Users don't play the game; instead, they program robots that play for them. The robots simply act out the consequences of their fixed rules. Each is equipped with simulated radar and weaponry to find and attack enemy robots, whose tactics are coded by other users. Perhaps future wars won't involve human lives but, rather, will be waged on the substrate of software programs grappling for an edge using game-theoretic tactics.

Warner's game was prescient: the early aughts saw a rapid rise in the use of uncrewed aerial vehicles as the US military targeted groups in the Middle East. Nations across the globe are still locked in a race to develop these technologies, even borrowing tactics from game-playing systems. In 2020, the US Air Force announced that it had repurposed MuZero, a board-game-playing algorithm created by Google Deep-Mind, to control the sensors of a U-2 aircraft for targeting and reconnaissance missions. This system, ARTUµ, highlights the US military's move toward fully autonomous decision-making systems, following game theorists' dream to relinquish onerous human responsibility to synthetic rationality. Mirroring this obsession with decision-making, Air Force executive Will Roper wrote: "The fact ARTUµ was in command was less about any particular mission than how completely our military must embrace AI to maintain the battlefield decision advantage."

Since the advent of Go and chess, military forces have tapped games to educate their soldiers. Kriegsspiel's data-driven scoring delivered games from the realm of abstract strategy to quantitative planning. In the 1980s, DARPA began funding video game companies to create games for training recruits. They continue to grant tens of millions of dollars annually to develop ever more realistic war simulations. In the early 1990s, they collaborated with game designers to improve flight simulators. This resulted in SIMNET, a multiuser flight simulator intended to train pilots. During the first Gulf War in Iraq, soldiers used SIMNET to test and rehearse complex tactical maneuvers. SIMNET gave rise to home programs and better video game simulations. Some of the hijackers in the 9/11 terrorist attacks had never flown real planes but had trained using Microsoft's *Flight Simulator* game.

The veil between war and games has continued to thin. McNamara tried to turn the Vietnam War into a numbers game, racking up a "score" of death and destruction. The first Gulf War was heavily televised.

Advances like night vision made it possible for civilians at home to follow battles on TV, distancing and abstracting the violence with an eerie green, video-game-like glow. This was the beginning of what cyberpunk author Bruce Sterling coined the "military-entertainment complex." Increasingly, soldiers are quarantined from firsthand violence behind computer monitors, controlling drones using repurposed Xbox controllers. "Players" are distanced from the moral consequences of their actions, their opponents disembodied and dehumanized by technological intermediaries. This distancing reduces—but doesn't eliminate—the risk of drone operators developing PTSD. Some military researchers have proposed that drone systems should incorporate anthropomorphized automated assistants that take orders from their operators. This way, the thinking goes, the operators would be less likely to feel that they themselves have committed destructive actions; it is a technological means of displacing guilt and assuaging their consciences.

Today, militaries across the globe use video games to recruit and train soldiers. The skills needed to excel at war have changed since the era of trench warfare, and the military is now explicitly recruiting gamers. In 2018, the US military founded its own e-sports team, and it has a Twitch channel on which it hosts giveaways that link to recruiting forms. Children as young as thirteen chat with recruiters on unmonitored channels. The US military's most successful current recruitment tool is the video game *America's Army*, first launched in 2002. Critics lauded the game for its realism: players learn how to dress wounds, coordinate with squad mates on tactical operations, and handle weapons. However, there is one notable discrepancy with reality: players don't die in *America's Army* without being immediately resurrected.

Other military organizations have followed the US's lead. Game developers in different countries responded to *America's Army* in kind.

In 2003, the Lebanese militant group Hezbollah released *Special Force*, which uses scenes involving the murder of children to evoke outrage in players and court recruits. A 2013 NSA leak from Edward Snowden revealed a concern that extremist groups were recruiting new members through online game forums. Members of the NSA, CIA, FBI, and Department of Defense created surveillance accounts to monitor players in online games like *World of Warcraft*. The games became so saturated with operatives that the agencies had to create a "deconfliction" group to coordinate their efforts after they realized they'd been wasting time surveilling one another.

These tactics are aimed at not only recruiting soldiers but also currying popular favor. Studies have indicated that games can alter the public opinion of military campaigns. By virtually experiencing acts of war, players become more likely to defend them. Former military and political officials staff top leadership positions at the game company Activision. Activision's hit game line, *Call of Duty*, features reimagined American military campaigns sanitized of moral ambiguity. Players realized that 2019's *Call of Duty: Modern Warfare* re-created an infamous Gulf War massacre along a road that subsequently came to be called the Highway of Death. In 1991, US troops and allies pummeled a retreating Iraqi convoy for hours, endangering hundreds of civilians, foreign workers, and surrendered soldiers in what Colin Powell called "wanton killing." The game, however, had Russian forces, not Americans, doing the killing. Many players were appalled by what appeared to be agitprop vilifying Russia for American misdeeds. Games have induced moral panic for millennia, from the dangers of gambling addiction to American Christians imagining "satanic" messages in Dungeons & Dragons. Today's panic usually focuses on video games' gore, but researchers have not found a link between in-game and real-world violence. Perhaps, instead, our moral concern should be directed at games' capacity to expiate the immensity of war's tragedy.

What has the toy lost compared with the real thing? Stakes. A child's miniature cash register does not employ a worker or feed their family. What looks like numbers on a screen for a trader to manipulate signifies the future of a retired couple. The video game soldier is shot and resurrected. The real soldier is not. We can pretend death and suffering do not exist in a game. We cannot do that in reality. Military strategists rushed to unburden themselves of the responsibility of decisions that could imperil humanity and delegated these decisions to game theory. Humanity narrowly escaped a handful of close calls, rescued by human—not synthetic—reason.

That the world's fate hung on glitchy detection systems is a staggering moral failure. The same is true today of our increasing reliance on automated surveillance and military systems, which are gaining mass adoption. In 2023, the European Parliament called for a society free from mass surveillance, banning the use of automated surveillance tools on its citizens. At the same time, it approved their use on migrants—the same logic by which Pope Innocent II banned the use of crossbows on Christians but sanctioned them for use against Muslims. Technology may successfully distance us from the horrors of war through gamelike interfaces, but it cannot obviate our moral responsibility to life.

The philosopher C. Thi Nguyen highlights a central aspect of play: though players locked in a game are nominally competing—even in a zero-sum game like chess—they are really cooperating to compete. Players agree to suspend rules of reality for a specific period, adhere to the rules of a game, and work together to achieve the shared goal of an enjoyable time. Competition is a means to an end: players must genuinely try to win to thoroughly enjoy themselves. "We might even call [games] a social technology, capable of converting aggression into a social benefit and perhaps even a moral good," Nguyen writes. This was Emanuel Lasker's original hope for creating a science of games, and von Neumann's hope for game theory. These were meant to be tools for minimizing violence. But their practitioners often leave out Nguyen's larger framing: the important insight that most people just want to get along.

The logic of game theory, in the abstract, is inarguable. Players are pulled inexorably toward equilibrium points. So, too, did mid-twentieth-century strategists believe themselves doomed to develop inevitable superweapons to maintain an equilibrium of power. Nations have since spiraled in place, investing trillions of dollars over the past century in ever more elaborate defense and missile systems instead of investing in positive-sum technologies. To remain a credible threat requires descending into madness and losing all moral scruples. The US military murdered hundreds of thousands of innocent people to underscore its credibility to other nuclear powers. Despite this, it's clear that our individual consciences aren't fully numbed. Until such atrocities can be fully automated—as by autonomous drones—researchers peddle in Band-Aid solutions to appease soldiers' consciences by distancing them from the consequences of their actions. All of this, in the end, might be seen as a project meant to off-load responsibility for difficult decisions onto machine systems—systems whose reasoning capacity, we're promised, will soon exceed that of any human. Yet this synthetic reason is nothing like ours, and we can't pretend that it's a valid or morally acceptable replacement.

Game theory was co-opted as the rational foundation of a neoliberal world order, and its selfish assumptions still ring through our engineered social systems. Originally invented to chart human behavior, game theory has since warped the territory it was intended to map. Whereas scholars studying games hoped to create a science that would end wars, the runaway proliferation of weapons has made us all more vulnerable. Historians have since discovered that America's nuclear race was, in many ways, one-sided. The German nuclear project, the original impetus for the Manhattan Project, was scuttled in 1942. RAND theorists were working with incorrect intelligence about Soviet atomic capabilities. They believed that the American military had fallen behind the Soviets, and this "missile gap" drove a near-messianic culture among analysts who were convinced they were saving the world. For decades, American strategists raced against their own imaginations. Today, the danger is more salient than ever. Nuclear diplomacy teeters on an even more precarious balance between nine nations. Some worry that a similar dynamic is playing out in the field of artificial intelligence as labs race to build a technology with potentially disastrous repercussions, convinced their technology could save the world.

As cryptographer Bruce Schneier argues, there's no such thing as security in the abstract. Antiviral software can't protect a user from the flu; a car alarm can't safeguard the vehicle's owner against identity theft. And sometimes security systems make us less safe because we grow complacent under their protection. Game theorists sought universal solutions in abstract mathematics, and the world is worse off for our leaders' faith in their technocratic solutions.

All of this should give us pause. In his novel *All the Names*, the Nobel Prize–winning author José Saramago writes: "Strictly speaking, we do not make decisions, decisions make us." If this is true, who are we when we relegate our decisions to mathematics? One of the hallmarks of humanity is how plastic and adaptable we are, as demonstrated by the diversity of human cultures and norms across the globe. The danger of using simplified models to inform political decisions is that humans learn. As the psychologist B. F. Skinner asserts, we can't harm or change physics by employing an incorrect model of atoms. But he was wrong to extend this to people: our models of humans can harm humans. Human behaviors are shaped by rewards and incentives, and our choices depend on what we believe about the games we play. Our beliefs about who we are can radically change the decisions we make and the way we live our lives. It's not a stretch to imagine that we might alter human dynamics by building social, political, or economic systems based on models incentivizing selfish behaviors. The reckless policies that drove nuclear brinkmanship suggest this to be true. Our theories about the world gave rise to profound moral slippage, a race to the bottom. Civilians were—and continue to be—held hostage by their country's military decisions.

Part III

Building Better Players

8

Chess, the *Drosophila* of Intelligence

In chess, a game of pure thought, unaffected by chance, wanting
to play against yourself is a logical absurdity. [. . .] Wanting to
play chess against yourself is as much of a paradox as jumping
over your own shadow.

STEFAN ZWEIG

The Austrian author Stefan Zweig and his young wife fled conti-
nental Europe in 1938 after the Nazis annexed Austria. They
pinballed between the UK and the US: Zweig wanted nothing
more than peace in which to work. Unlike his fellow writer in exile
Thomas Mann, he could not bring himself to be openly critical of the
Nazis. He wanted only to be left alone. Nowhere in Europe seemed
safe, so Zweig and his wife moved to Petrópolis, Brazil, in 1941. Here,
he effuses, "away from politics," one could live "closer to one's own and
nature's heart." Japan's bombing of Pearl Harbor set him into a panic,
and he grew terrified that the Axis powers would conquer even the
Americas. He felt increasingly alienated from his life. Zweig was isolated in a foreign land with his equally depressed wife, cut off from
former friendships, "miles and miles away from all that was formerly
my life, books, concerts, friends and conversation."

Zweig's final novella, *Chess Story*, reflects this experience. The book follows Dr. B, an Austrian financial adviser with ties to the Catholic Church and the fallen aristocracy. Dr. B is arrested in Vienna and tortured by Gestapo agents hoping to seize his clients' fortunes. Imprisoned in an empty hotel room for months, he is deprived of human interaction, books, and writing materials: "There was nothing to do, nothing to hear, nothing to see, everywhere and always there was nothing around you, a complete timeless and spaceless void. [. . .] You remained alone. Alone. Alone." He manages to pilfer a guard's pocket, stealing a book on the games of chess masters. After memorizing the 150 games in the book, he discovers that he can play chess games against himself in his mind, splitting his psyche into an "I white" and "I black." This distraction allows him to resist the Gestapo's torture tactics for a while, but he plays so obsessively that he suffers from "chess poisoning" and has a nervous breakdown. The Gestapo releases him once they realize that he can't provide useful information in his state of insanity. The mental realm had offered a temporary liberation from prison until it caged him within the strictures of the game.

Dr. B eventually recovers and escapes Europe on a ship bound for Argentina. Here he meets Czentovic, then the world chess champion, who plays well, though mechanically. Czentovic is motivated by money alone, uninterested in the intellectual aspects of the game. Unable to read or write, he's thought to be a stand-in for the uneducated fascists who nevertheless excelled in rules and regulations. Other passengers goad Dr. B into playing the boorish champion. Dr. B wins the first game and is doing well in the second until Czentovic discovers that he can agitate his opponent by drawing out each move. Each pause leaves Dr. B alone with his thoughts, as in that Viennese hotel room. His mind spins out, making frantic internal calculations while manically playing against itself. Dr. B eventually chooses to resign rather than descend into madness. He vows never to touch chess again.

Zweig wrote *Chess Story* in 1941, the year before he and his wife died by suicide. Like Dr. B, they'd decided that the only way to win was not to play. Zweig's biographer, George Prochnik, describes the police photo of the scene: Zweig's hands are crossed, his wife's head on his shoulder, her hand over his. "He looks dead," Prochnik writes. "She looks in love."

Chess Story depicts a mind transcending desolate circumstances by inventing another player. With it, Zweig anticipated what would become one of the great endeavors of the last century: the attempt to re-create intelligence in silico. Games would become central to this quest, as computer scientists focused on simulating the purposeful nature of intelligence. Chess came to be known as the *Drosophila* of AI: like the geneticist's favorite model organism, the fruit fly, chess became the standard model against which engineers measured their programs' intelligence. Beyond this, engineers used games as the playground on which their programs learned and bootstrapped their capabilities. Much as Dr. B pits his intellect against itself, engineers pit programs against copies of themselves in self-play. These programs learn to adapt to the demands of the game, contest by contest, rather like the churn of organisms adapting to their environment over the course of evolution.

Something magical happens when an agent is endowed with a desire and the basic means to achieve it. Living organisms want to survive and reproduce. Games, instilled with inherent goals, are an excellent medium for mimicking this. The study of games—both by players immersed in them and researchers at a remove—reveals the logic of desire. Von Neumann invented a mathematical technique that can anticipate an agent's choices simply by knowing what they want. He believed that human selfishness was a "force of nature." The conservation of energy makes it possible to predict the motion of a physical system. Similarly, von Neumann believed self-interest governs human activity and renders people explicable. This notion was not

new. In Indian philosophy, suffering arises from humans' unquench-
able appetites. Desire is the unifying principle beneath the multiplic-
ity of appearances. Striving reveals itself as peace once its true nature
is recognized: it has no end and therefore nowhere to go. Desire is both
the source of endless suffering and its ever-present cure. The Greek
poet Sappho characterized desire as a "bittersweet creature against
which nothing can be done." Plato spoke of conflicting desires: though
people fundamentally seek truth, this goal is thwarted by the body,
which "fills us with wants, desires, fears, all sorts of illusions and much
nonsense, so that, as it is said, in truth and in fact no thought of any
kind ever comes to us from the body." Bodily desires, he argued, are at
the root of all war and civil discord. The seventeenth-century philosopher Baruch Spinoza contended that "desire is the very essence of a
man." The early exponents of probability theory proposed that people
make decisions to maximize their rewards. Capitalism was founded
on the belief that desire actuates economic output. Freud believed
that people are unconsciously driven by an unbounded torrent of sexual desire. The German philosopher Arthur Schopenhauer, heavily
influenced by Indian thought, believed that the universe is one uni-
tary, senseless striving; will is the essential reality veiled by the world
of appearances.

Desire is the first ingredient of agency. Endow something that can
take actions with a desire—whether acquired over evolution or installed
by programming—and you have the simplest recipe for something re-
sembling intelligence. Complex, goal-directed behaviors emerged over
the course of evolution in service of the original aim, survival. Paleobi-
ologist Douglas Erwin writes:

Paleontologists often say that a burst of diversity in the fossil record
simply "filled in ecological space," as if each new species simply
took up residence in a square of a preexisting chessboard. [. . .] I

think a much better analogy is of building the chessboard itself. Although some of these ecological spaces may exist independently of any species that occupies them, many more are defined by species and their mutual interactions.

Evolving animals are not only adapting to their sterile, static environments—they're adapting to one another. New species may change existing niches or become niches themselves. Two billion years ago, photosynthetic algae radically altered the ancient atmosphere, providing an oxygen-rich milieu that fueled the emergence of more complex life forms. Coral reefs became the home of thousands of aquatic species. Land-based lichens and fungi prepared the rocky earth for multicellular plant and animal life. This dynamic is true of not just physical niches but also behavioral adaptations. With the evolution of the nervous system, animals acquired the ability to move purposefully, bootstrapping increasingly intelligent strategies and countermeasures. Once a predator gained the ability to move and hunt prey, other animal populations also profited from gaining the ability to move. New behaviors and strategies emerged to counter those of other players, eventually resulting in a runaway intelligence explosion that led to modern humans.

The complexity of an environment limits the level of intelligence that its inhabitants can attain. Human-level intelligence wouldn't emerge in a world populated by only bacteria—it would be overkill. Likewise, the complexity of a game limits what its players can learn. One can be a genius chess player, but there's no such thing as a tic-tac-toe prodigy. AI researcher Joel Leibo and his colleagues imagine a single "solipsist" learning, by trial and error, a game of solitaire on a nineteen-by-nineteen grid. The player's goal is to place stones on the board to capture as much territory as possible. "Obviously, the optimal solution is simply to place stones all along the edge of the grid." Once the player discovers this, there is nothing more to learn. The agent's intelligence hits a ceiling and stagnates. Now, introduce another player who can block the first player's moves by placing white stones. "The white stones and the territories they enclose become barriers, preventing additional expansion of the black territory. Thus, the game of Go is born—a game with enough emergent complexity to occupy millions of minds for millennia."

In their quest to re-create intelligence from scratch, some researchers have begun taking lessons from its original source: life. Biological intelligence, it's becoming increasingly clear, emerged thanks to cooperation as well as competition. Early life, consisting of a sea of undifferentiated protobacteria, didn't evolve according to the traditional model of competitive Darwinism. There weren't necessarily species per se. Genes were not only passed down to offspring in vertical transmission. These bacteria-like life forms shared evolutionary innovations—new genes and mechanisms—with one another through a process called horizontal gene transfer. Like modern bacteria, they collaborated in vast, protective microbial mats. Another cooperative insight paved the way for new, more complex life forms. One unicellular cousin took refuge inside another, forming the collaborative venture now known as the eukaryotic cell. Their internalized mitochondria provided so much energy that eukaryotic cells could perform entirely new functions, specialize, and work together to make up multicellular animals. This eventually gave rise to the Darwinian form of evolution we know today, with its hereditary transmission of genetic material and elaborate speciation.

As biologists and AI researchers alike realized, the process of evolution is a form of learning. It tunes the gene pool, which orchestrates a cell's protein machinery, to the requirements of its environment. Different organisms discovered distinct solutions to the problems posed by their surroundings. Some species are successful because they specialize in a specific niche. The yeti crab, for instance, is confined to shuffle within a few square meters around deep-sea water vents in Antarctica, thriving in water temperatures up to seven hundred degrees Fahrenheit. Other species became generalists, evolving ways to deal with environmental variability. Mammals' warm-bloodedness, for instance, enabled them to migrate to colder climates and operate at night. Major environmental upsets, like volcanic eruptions and asteroid impacts, caused mass extinctions in hyperspecialized species, favoring generalists that could better handle disruption.

One way of dealing with environmental change is known as phenotypic plasticity: changing one's body shape, for instance, in response to environmental stressors. The water flea grows protective spikes during development if it senses high levels of chemicals secreted from predators in its home waters. The migratory locust has a solitary form that looks much like a grasshopper's. If they feel crowded by other locusts—suggesting that there might be more locusts than available food—this triggers a morphological switch. They metamorphize into a gregarious form: the biblical plagues of locusts, who band together into massive swarms and devour crops. Different castes of honeybees, such as workers and queens, develop depending on the amount and quality of food that nurse bees give larvae. In laboratory settings, a larva given an intermediate amount of food develops a body type halfway between worker and queen.

In phenotypic plasticity, an environmental cue sets off a change in the way an organism's genome is read, altering how its body and behavior are expressed. Interestingly, in all these examples, the switch from one form to another is mediated (at least in part) by dopamine, which we know is also responsible for another way that animals handle environmental variability: learning. Dopamine is by no means responsible for all phenotypic plasticity in all species, but it's been repurposed to manage variation in both body plan and behavior. Bodies and behaviors are expressed in concert with the local environment and its inhabitants, a great game orchestrated by historical contingencies meeting present-day needs. Both behavioral and bodily plasticity evolved to meet environments changing at timescales faster than evolutionary changes to the genome. Perhaps an ecosystem becomes dramatically wetter; an animal might adapt by learning to build an insulated nest. Animals can discover novel behaviors through play. Play helps the nervous system generate diverse behaviors from which to select new actions. It allows animals to simulate experiences and interactions with their peers, a form of safe exploration.

Random variation is a search logic that has been discovered again and again. Evolution progresses along genetic mutations, which produce variations for natural selection to choose from. This variation can give rise to bodily diversity, allowing animals to express various forms in response to their environments. Learning and play produce behavioral variability. An animal's nervous system can learn and store a repertoire of behaviors to select from in future circumstances. Play amplifies an animal's agency by expanding this repertoire of behaviors. In the same way that stunning morphological complexity has been bootstrapped from random mutations, intelligence has been bootstrapped over billions of years of spontaneous play. The basic algorithm is this: explore many options through variation, catalog what works (whether through survival of the fittest or memorization of successful strategies), and put what works into practice.

Life is not a single-player game, of course, and animals evolved intelligence as much to deal with other players as to deal with their environments. For centuries, scientists had taken for granted that brains evolved to process information about the world. They thought that the goal of the nervous system was to represent the environment faithfully. Accordingly, neuroscientists focused their research on sensory processing: visual perception, audition, touch. By the 1970s, biologists realized that some animals may have developed large brains to better organize their complex social bonds. Building and maintaining relationships requires intense cognitive effort, running the gamut from communication to deception. In keeping with the ethos of the time, scientists initially dubbed this the Machiavellian hypothesis, suggesting that primates evolved intelligence to manipulate one another.

British anthropologist Robin Dunbar eventually renamed it the social brain hypothesis. He'd discovered that primate brain size tends to scale with their average group size. Primates presumably needed bigger brains to maintain bigger social groups, and bigger groups tendered survival benefits. In groups, primates can collectively protect against predators, share parenting or foraging duties, and innovate on cultural technologies like tool use. Dunbar used the average human brain size to extrapolate an expected group size for humans and found it to be about 150. This is the origin of the oft-repeated claim that human groups max out at around 150 members, offered as though this were an incontrovertible limit on human empathy. But this is not a data point; it is an inference—and a highly contested one with massive error bars. Dunbar's number is sometimes used to explain why humans naturally form in and out groups, or to suggest that racism is inescapable because people can only care about 150 other people. This extrapolation is evidence of no such thing.

The social brain hypothesis isn't a universal truth of biology. Social complexity isn't the sole driver of the evolution of bigger brains. An individual from a social wasp species is not necessarily more intelligent than one from a solitary wasp species. Colony-forming species may, in fact, lose brain complexity as individuals become narrowly specialized for given hive tasks. Wasps recognize their colony mates by the scent of their shared pheromones, not by memorizing the faces of hundreds of other wasps.

When Dunbar compared the brains of other mammals and birds to their social behaviors, he found a surprising difference. Outside primates, brain size doesn't correlate with group size. Instead, animals that mate monogamously have the biggest brains. Perhaps it wasn't the computational demands of living in large groups that sparked higher intelligence but the demands of pair-bonding. Primates, then, reappropriated the social skills originally developed to connect with their sexual mates to build relationships with other individuals, such as friendships. Perhaps higher intelligence then grew from this platform of collaboration. Scholars have discovered that teams of diverse thinkers perform better on difficult problems and come up with more creative solutions than individuals or homogenous teams. AI researchers have begun adopting this insight to train more robust and collaborative agents.

Life—and intelligence—emerged from interaction. In Zweig's *Chess Story*, Dr. B transcends the torture of solitary confinement by inventing another player in his mind. Darwinian evolution likewise proceeded through the invention of new players. Each honed their adaptations against the blades of one another. Every organism owes what it is to other life. Natural selection gave rise to a diversity of body plans adapted to specific lifestyles and environments. It also invented mechanisms for those body plans themselves to change depending on their environments. With the evolution of the nervous system, animals could react even more rapidly to novel stressors and challenges. After billions of years, human intelligence emerged, bootstrapped by social interplay. Engineers hoping to re-create intelligence in silico took note of biology's simple recipe: desire plus interaction. Games would provide an ideal test bed for their efforts. Whereas living organisms are endowed with the desire to survive, artificial agents are endowed by researchers with a desire to win.

9
The End of Evolution

What is life? What is it that enables living things, apparently so
moist, fragile, and evanescent, to persist while towering
mountains dissolve into dust, and the very continents and oceans
dance into oblivion and back?

ROBERT ROSEN

efore Darwin, the natural world had been thought to be strung in
a balance so perfect that its very details could be lifted up as proof
of the existence of God. Much as people had believed that dice
rolls revealed the divine will, they thought the natural world reflected
God's careful design. Every creature was perfectly adapted to its niche,
as if intricately crafted to specification. In 1802, the theologian and
utilitarian philosopher William Paley published his last book, *Natural
Theology*, which rehashes this argument with extensively researched
biological examples. The fins of fish and wings of birds are superbly
suited to the water and air in which they operate, he notes. A salmon's
inborn nostalgia drives it back to its birthplace to safely spawn. The
crossbill's divaricate beak is an ingenious mechanism for extracting
the seeds from pine cones. Paley's simple incantation of biological details read, at the time, like self-evident proof of a Creator, each impossibly perfect correspondence announcing His infinite benevolence.

Should one pitch one's foot against a stone in the forest, Paley argues, one will hardly wonder how it came to be there. But if their foot hits a pocket watch instead, they will have to admit that it has been placed there, that someone made it. A watch isn't accidental, like an errant pebble. It has custom-built mechanisms that work together; it has a clear purpose. It has to have been created by someone.

Charles Darwin read Paley as a university student and found his arguments convincing, though he didn't think about them too deeply. He confessed he'd been as charmed by Paley's logic as by Euclid's lucid geometry. But Darwin's fieldwork would later come to dispel that enchantment. Natural laws can give rise to the appearance of design, shaped by environmental necessity. In conceiving his theory of natural selection, Darwin was inspired by the work of Thomas Malthus. In his 1798 book, *An Essay on the Principle of Population*, Malthus contends that progress for mankind is ultimately Sisyphean. Improving material conditions in a human society inevitably increases the population, thereby lowering conditions through famine and disease. Darwin saw in Malthus's writing a parallel with nature, a "struggle for existence" that gave rise to new species. Life, Darwin reasoned, is not contrived by a Creator, whose hand steadies His kingdom. Life is not stable at all—it's in flux, though driven by fixed laws. The elements of nature are shaped by forces balanced in opposition. All that is arises out of conflict, the true law of nature.

Darwin took decades to publish his theory, knowing it would shock society. It did, and continues to. Darwin sent his work to the astronomer John Herschel, one of his intellectual heroes, but learned through a roundabout channel that Herschel was contemptuous of the theory, reportedly calling it "the law of higgledy-piggledy." As the Greeks had overlooked chance in their mathematics, scientists remained blind to the power of randomness. Accident could not possibly play a role in the harmony of nature. How could creation be reduced to the roll of dice?

Chance had previously been used in service of another long-standing argument thought to prove God. Naturalists had determined that the ratio of males to females was balanced in many species, which seemed to indicate an intentional design. By the eighteenth century, England had begun collecting large social datasets that offered new evidence in favor of this argument. In 1710, the mathematician John Arbuthnot published his analysis of English christening records. He'd discovered that males and females were not born in a neat one-to-one ratio. Instead, there were about fourteen boys born for every thirteen girls. Strangely, this had held true every year for the past eighty-two years. Arbuthnot took this as evidence of God's foresight. Males, he reasoned, were subject to more "external accidents" (for example, exposed to danger when hunting), resulting in more male deaths. The initial surplus of males was tempered by their more adventurous nature, such that, once they reached reproductive maturity, a perfect one-to-one ratio of men to women was maintained. This, Arbuthnot argued, could not be due to chance. Assuming that the default should be an equal ratio of males to females, the probability that males happened to exceed females, by chance, for eighty-two years straight was vanishingly small. He took this as irrefutable evidence of divine design. For every person, God provided a spouse.

Darwin offered another solution to the mystery, one at odds with his own theory: group selection. He'd initially hypothesized that natural selection works at the level of competing individuals. Perhaps, he surmised, there are some cases in which natural selection works on an entire population. This would explain the emergence of adaptations that don't necessarily benefit individual animals but confer benefits to the group as a whole. How else could one explain social insects like ants, whose colonies comprise sterile individuals toiling for their mother and queen? These individuals apparently eschew reproduction for the good of their species. The idea that some animals work together for the greater good was a natural intuition. Centuries earlier, philosopher Thomas Hobbes had written that for ants, unlike humans, "the common good differeth not from the private." Darwin realized, however, that this didn't quite add up. How would an individual ant know what's "best" for the group? What compels it to serve its queen? He suspected that the balance of the sexes must somehow be advantageous for the species as a whole, "but I now see that the whole problem is so intricate that it is safer to leave its solution to the future."

No researcher could work out how natural selection acted on successive generations until scientists knew what, exactly, the substrate of inheritance was. Gregor Mendel, in his careful breeding of pea plants, was the first scientist to realize the statistical nature of evolution. The twentieth-century mathematician Ronald Fisher would develop this insight more fully. Whereas Mendel had sketched the rough rules orchestrating inheritance, Fisher worked out how this related to the broader flow of evolution, describing natural selection mathematically using the logic of chance. Fisher was a great believer in the power of randomness and leveraged it to improve mathematical techniques and experimental designs. He intuited the many ways in which chance is productive. Evolution, for instance, leverages randomness, recombining and shuffling alleles to sieve out the most successful mash-ups in a manner that yields miraculous-seeming forms. As biologist Julian Huxley puts it: "Natural selection plus time is a mechanism for generating an exceedingly high degree of improbability." Exquisitely organized animals arise from random happenstance, with minute changes compounding over generations.

Though it came decades before the discovery of DNA and its mechanisms, Fisher's work in genetics was hugely clarifying for evolutionary biology. He characterized evolution as the ebb and flow of gene variants in a population. A gene that influenced body hair could have multiple variants, or alleles: one specifying thick body hair, another specifying no body hair, and so on. If one of these alleles had even a tiny effect on reproductive success—an ice age arrives, say, and body hair helps an organism stay warm—the allele coding for thicker body hair would become more common in successive generations. Should circumstances change—the ice age ends—making that trait variant maladaptive, it would become rarer in the population. Fisher boiled evolution down to statistical flows. What was once thought to reflect God's design could be characterized with the same mathematics that govern dice.

In 1930, Fisher collected and extended his genetic insights in his opus, *The Genetical Theory of Natural Selection*, which almost single-handedly revitalized Darwin's ideas and rescued them from years of neglect by academics. In the book, Fisher elaborates—in his characteristically cryptic style—what is now one of the most celebrated arguments in evolutionary biology and an early seed of game-theoretic thinking in the life sciences. Fisher revisits the question of the balance of sexes and resolves the mystery of how it can be enforced by natural selection without resorting to muddy ideas involving group selection. To be fair, Darwin and other thinkers had anticipated his argument, but Fisher expresses it with greater clarity.

Say that a population is biased to have more female than male offspring, resulting in an unbalanced sex ratio. In the next generation, males, being a smaller proportion of the total population, will have more partners to choose from and so will have relatively more offspring. Any propensity to have more male offspring will be disproportionately passed on to the next generation, leading to more male births and rebalancing the male-female ratio. Imbalances will be quickly corrected in successive generations, because the rarer sex will necessarily have more reproductive success. This became known as Fisher's principle, describing, in essence, a game between the sexes.

What, then, of Arbuthnot's proof of God, citing the dramatic unlikelihood of nature providing fourteen men for every thirteen women? Fisher realized that this game isn't simply about maintaining a one-to-one ratio. The balance depends on how much work parents must put into their offspring. Nature will find its equilibrium point wherever the total effort required for producing offspring of either sex is equal. Given that male children die slightly more often than female children, they require less parental investment on average, so natural selection had tipped the scale in their favor. The stable sex ratio was not proof of a Creator but rather the equilibrium point of a game.

By the turn of the twentieth century, Darwin's shocking theory had become more widely accepted in the scientific community. These ideas, taken entirely out of context and warped to justify political ideology, were used to rationalize some of the century's greatest atrocities. Francis Galton, Darwin's cousin, coined the term *eugenics*, defining it as "the study of the agencies under social control that may improve or impair the racial qualities of future generations, either physically or mentally." He would go on to establish the Galton Laboratory for National Eugenics at what is now known as University College London. Several famous scientists and public figures were vocal eugenicists, including Fisher, H. G. Wells, and Winston Churchill. Many of them were not proponents of sterilizations (and certainly not death camps) but, rather, of "positive" eugenics—for example, programs offering tax incentives for "desirable" (that is, upper class) people to have more children.

Eugenicists believed that their "advanced technology" would cure all social ills, from alcoholism to crime. It would instead reanimate an ancient horror. German scientists had been among the earliest and most enthusiastic adopters of Darwin's ideas. But Americans were the first to establish a eugenics program in practice. Its proponents saw it as a rational, scientific solution to the problems of a complexifying society. Over thirty American states enforced compulsory sterilizations. Many of their victims are still alive today. Unfortunately, the specter of eugenics continues to haunt modern discourse, from outright racist ideologues to Silicon Valley start-ups promising (scientifically impossible) positive eugenics–based methods of embryo selection. Its proponents envision humans as metrics to be optimized, while failing to recognize that diversity is the substrate of evolution. Narrow uniformity spells existential peril for any population. In this imagined game whose score is measured by hypothetical fitness, morality is marginalized: an ornament of a functioning society, not its bedrock.

The utopian impulse of eugenics was based on the false premise that genes, as opposed to the environment, are the primary source of human qualities and behaviors. Fisher believed that all of a person's virtues and vices—their physical beauty, moral instincts, religious awe—were dictated by their genes. Eugenicists embraced this notion, arguing that humanity might be "purified" through selective breeding and culling. As the genocidal ramifications of eugenics ravaged Europe, Fisher and many others distanced themselves from the philosophy. Eugenics-focused research journals and laboratories rebranded themselves with the sanitized moniker "genetic." The shadow of Nazism continued to loom over biology for decades, with chilling research that even tenuously suggested that heredity plays a role in human behavior.

Despite the prevailing intellectual climate of the 1960s, graduate student William Donald Hamilton was determined to investigate the genetic basis of altruism. As an undergraduate student at Cambridge, Hamilton taught himself biostatistics by poring over Fisher's book. Hamilton began to see all living things as being poised at the equilibrium of a great game. Organisms churn through the cycle of birth and death, contending for the grand prize: the opportunity for their off-spring to play in the next round. These games—between species, within species, with the environment, between genes—define the equilibrium forces balancing the networks of life. He was particularly drawn to the mysterious emergence of altruism, seemingly incompatible with Darwinism. If life is a struggle between individuals for survival, how could selflessness ever emerge? Hamilton believed there must be a better explanation than group selection.

In 1963 and 1964, Hamilton published two papers on social behaviors, pioneering the "gene's-eye view" of evolution (a framing that would later come to be popularized, more confusingly, as the "selfish gene"). In this framework, evolution is a battle between competing genes instead of competing organisms. How could a gene for an altruistic behavior arise in a population, given that altruism can be costly—sometimes even costing an individual their life? From a gene's perspective, it doesn't matter if an altruistic individual carrying it dies, so long as other copies of that gene survive in the group. Hamilton used the example of birds alerting their neighbors to the presence of a hawk. The bird raising the alarm incurs a small extra risk because it makes its position known to the predator. But that cost, from the gene's perspective, must be worth the benefit of alerting nearby birds, so long as some of them also carry the same gene. He derived a simple equation to describe when a gene for an altruistic behavior could arise. The cost of the behavior must be less than its benefit to other copies of itself in the population. Altruism could be thought of as an economic calculation, which Hamilton pithily expressed in his paper: "In the world of our model organisms [. . .] everyone will sacrifice [his life] when he can thereby save more than two brothers, or four half brothers, or eight first cousins." Hamilton's idea came to be known as kin selection, in contrast with group selection. These behaviors tend to favor closer relatives more than distant relatives, since closer relatives are more likely to share gene variants. The rule doesn't require that an altruistic gene "recognize" itself in another individual—the gene might influence behavior according to some blanket rule, like "feed any open mouth I see, but only if it's inside my own nest."

Hamilton's work would soon catch the eye of the academic dilettante George Price. In 1967, at the age of forty-five, after a life that already seemed like a chronic midlife crisis, Price boarded a boat for London. He and his wife had divorced over a decade ago; he hadn't seen his two daughters in years. A botched thyroid surgery performed by an old friend had left George partially paralyzed, in pain, and seething with regret. "May go haywire," an interviewer noted, decades earlier, on George Price's 1940 Harvard application. Price won a coveted scholarship despite his "baffling" disposition. His career was one great frantic digression, pinballing between major technological breakthroughs of the era. During his graduate research, Price was involved with the Manhattan Project, working to characterize plutonium-235. After a short stint as a lecturer at Harvard, he joined Bell Labs to improve the chemical properties of transistors, just then emerging as the driving force of the computing revolution. Later, he worked on a game-theoretic analysis of Cold War military strategy he'd never finish. Meanwhile, he approached IBM with an idea for a universal "Design Machine" that presaged CAD programs. He worked at IBM briefly before tiring of the project. Through the decades, Price published erratically, but his mind never stopped casting about for a worthier cause, desperate to make his mark in science.

Price had lied to his daughters about his London trip. In a letter, he claimed he was only planning to visit for a few months to write some magazine articles. To a colleague, he admitted the truth: he planned to use the insurance settlement from his failed surgery to support his research into the origin of the human family. If that didn't work out, he was not opposed to ending things. Perhaps motivated, unconsciously, by his failure to provide for his own family—he chronically shirked paying child support—Price had grown fascinated by the origin of paternal care. Few other male mammals are involved in child-rearing. What makes humans different? He spent his days padding through London's sprawling network of libraries, teaching himself evolutionary biology from scratch. He was arrested by Hamilton's paper reducing altruism to nepotism, though it struck him as too cynical. What of the countless stories of human generosity to absolute strangers?

Price struck up a correspondence with Hamilton, who was off doing fieldwork in Brazil at the time. Hamilton replied with a copy of his most recent paper, in which he'd used game theory to extend Fisher's argument about sex ratios. Not all species have an equal sex ratio. A female of the parasitic wasp *Mellitobia acasta*, for example, lays a brood of approximately fifty eggs in the body of a bee pupa. Only one of these is male. After they hatch, the wingless male inseminates all his sisters before they consume the pupa and fly off to lay their own eggs, leaving their brother to die. Hamilton used game theory to model several species with nonstandard mating and sex-determination mechanisms. He ran computer simulations wherein the male- and female-determining genes were players and fitness was their payoff. The players stabilized at specific sex ratios, which fit remarkably well with the data. Hamilton's game-theoretic models predicted the rather bizarre sex ratio of insects, from *Anaphoidea nitens* (with one male for every three females) to *Siteroptes graminum* (with one male for every twenty females). Hamilton called this the species' "unbeatable" strategy, akin to a Nash equilibrium, from which no player would want to deviate.

Much as chance had once seemed lawless, much as economics had seemed recalcitrant to mathematics, biology had long been thought impenetrable to systematization, then yielded to the logic of games. Researchers adopted game theory to describe the process of evolution and predict its end points. Price was fascinated by the idea of modeling evolution as a game—he had, after all, spent years contemplating game-theoretic nuclear deterrence strategies. Key to the delicate balance of peace was not actually using nuclear weapons—this would end in mutual destruction—but blustering a sufficiently credible threat that one *would* use them. Similarly, Price realized that, though conflict is almost universal across nature, the animal kingdom is rife with examples of limited, rather than total, war. Antlers, for instance, had long baffled naturalists, who couldn't quite work out their purpose. They seemed a mysteriously costly ornament or an inexplicably ineffective weapon, if they were even meant for that. They're too blunt to inflict mortal damage. Wouldn't it have made more sense for deer to have evolved sharp horns with which they could murder their rivals during mating season? Instead, rutting bucks use their antlers to wrestle other males for mates in a relatively harmless display of strength. So where was all that bloodlust of life, supposedly red in tooth and claw? Why was nature content to commerce in genteel play, like plumage displays and ceremonial battles?

The ubiquity of limited conflict was, at the time, thought to be further evidence of group selection: nature putting the "good of the species" above the individual and forestalling mass murder every mating season. Putting this notion to the test, Price used game theory to simulate repeated fights between individuals. He modeled a population of individuals with varying propensity to de-escalate conflict. Price iterated these simulated populations over several generations, discovering that animals willing to de-escalate fared best. Limited war benefited both individuals and the group. Even if aberrant animals with deadly antlers evolved, they would not take over the population. The limited combat strategy always dominated. The word *strategy* here does not imply that these are reasoned-out decisions. They are, instead, genetically specified behaviors or adaptations.

Conflict, Price concluded, could also be harnessed in service of the collective. African hunting dogs, a highly cooperative species that hunts in packs, will sometimes mob a noncooperative member who hasn't been pulling their weight during hunts. This kind of policing might be a mechanism to enforce cooperation in a prisoner's dilemma. Shirkers are punished, and the punishers avoid being punished themselves. Unrelated individuals can still behave altruistically—their altruism need only be sufficiently beneficial to the group and actively enforced.

Price's paper was accepted in principle by the journal *Nature*, but he would never get around to making the revisions needed to publish it. It was discovered among his files decades later by the historian Oren Harman. Nevertheless, its reviewer, John Maynard Smith, was taken by Price's idea. He reached out to Price to collaborate on a more mathematically rigorous treatment of the problem. In 1973, they published "The Logic of Animal Conflict," a paper that would spawn a new field.

In it, they imagine a species in conflict over a limited resource. Individuals can adopt different strategies, commonly referred to as "hawk" and "dove."* Individuals using the hawk strategy initiate fights whenever they encounter another of their species. Doves, on the other hand, run from conflict, but they will share resources equally if no fight ensues. Which strategy will dominate the population over the course of successive generations? If a hawk meets a dove, the dove will back off, and the hawk will get the entire resource. If a dove meets a dove, they will split the resource fairly in half. If a hawk meets a hawk, they will fight. On average, each hawk will win half the resource, minus the cost of fighting. While it might appear that the belligerent hawks should dominate the population, the dove's pacifist strategy is

---

* Price and Maynard Smith dubbed these strategies "hawk" and "mouse" because Price, who'd recently become devoutly Christian, did not want to sully the biblical symbol of the dove.

more resource efficient. This results in a mix of both populations. Suddenly, a wide range of previously unexplainable biological phenomena became interpretable in light of game theory. Striking behavioral diversity can coexist within a single species. Price and Maynard Smith deemed these evolutionarily stable strategies, akin to Hamilton's unbeatable strategy and Nash's equilibrium. As long as the environment doesn't change, other strategies can't do better than an evolutionarily stable strategy.

To fit game theory to biology, Maynard Smith, Price, and others had to loosen its classical assumptions, which posit rational players with impressive foresight. Bacteria don't have minds but come equipped with remarkable strategies. Some manufacture poisons and antidotes to engage in chemical warfare with their competitors, while others have protein lances to invade their neighbors. Still others cluster in protective biofilms. In evolutionary game theory, a player can be anything with an aim—a computer, a corporation, or a cancer cell. Organisms don't need brains to play games—natural selection is the decision-maker, optimizing for fitness instead of points. The mindlessness that evolutionary biologists introduced to game theory would later prove helpful in engineers' attempts to create artificial intelligence.

Richard Dawkins later popularized the framing developed by Hamilton, Maynard Smith, Price, and others in his 1976 book, *The Selfish Gene*. The same argument had been put forward a decade earlier by George Christopher Williams in his book *Adaptation and Natural Selection*. However, this failed to attract as broad an audience as Dawkins's more bombastic treatment. Genes, they both argue, are the fundamental unit of selection. Genes are what play the game of life, and we, the individual manifestations of those genes, are a sideshow. These books popularized a promising theoretical framework, though one not universally accepted among biologists. The gene's-eye view of evolution employs a hazy definition of genes, arguably gaining a handy explanatory metaphor at the expense of genuine biological meaning-fulness. Dawkins later argued that the selfish-gene framing should be thought of like a Necker cube illusion: one way of seeing biology, and not necessarily more accurate than other views.

Dawkins appropriated the value-laden word *selfish* to define a technical term—much like game theorists had appropriated *rational*. A gene's selfishness is very different from that of humans. Genes don't "want" a particular outcome. This confused many readers, and sometimes even Dawkins—particularly in the book's first edition, where he conflated genes with people, professing that "we are born selfish." Despite disclaimers disavowing genetic determinism, it's hard to come away from the book without some sense that genes are purposefully selfish puppeteers of the bodies that house them. It's a common cognitive bias, reflecting the same innate generosity with which a child endows their teddy bear with consciousness. Humans cannot help projecting intention on inanimate entities. It's almost impossible to scrub out this impulse entirely, even when using the most careful language. Dawkins later regretted choosing the term *selfish gene* and wondered if *immortal gene* might have been a better choice. These ideas have been part of the popular conversation for nearly fifty years, yet many readers and academics still confuse mathematical models of genes with the behaviors of real people.

Today, the gene's-eye view of evolution is considered one of game theory's most celebrated successes. It's also procrustean, distilling life's richness into simple equations. While it's easy to criticize game theory for oversimplifying complex dynamics, that's also the point: all models are simplifications. The physiologist Denis Noble has argued against this genetic reductionism in biology. There is no genetic code for a cell's lipid membrane; an embryo inherits mitochondria and patterning signals directly from its mother. Life can't be fully reduced to the alphabet of nucleotides constituting the genetic code. "The book of life," he writes, "is life itself."

Despite its limitations, game theory has made biology more of a predictive science. Cancer medicine is a promising example of this. Cancer is evolution gone awry, the revelation that the cells comprising us are still capable of independence. For multicellular life to work, nature found ways of restricting competition between the cells of an individual. Cells are usually law-abiding citizens of their bodies. They divide only as needed to function. Cancer cells are different. In dividing, a cell might accrue some mutation that takes all the brakes off replication. The cells now divide out of control, accruing new mutations faster, behaving more like selfish individuals than a cooperative whole. Cells usually have extensive error-checking mechanisms to ensure that their genetic material is carefully copied, but these may also be disabled in the frenzy of growth. By way of mathematical inevitability, any mutation that allows a cell to divide more rapidly will spread in the population of tumor cells. With their error-checking mechanisms disabled, these rogue cells accrue mutations ever faster and often gain new functions. Some regress to the ancient habits of their undomesticated, unicellular ancestors, such as reverting to fermentation. Some cancer cells have a kind of unicellular sex, allowing them to further mix up their genes. Others gain the ability to bud off and circulate in the bloodstream to colonize the rest of the patient's body.

Oncologists aim to match chemotherapy drugs to specific cancers, but cancer itself is a moving target, as most tumors are a heterogeneous mix of rapidly mutating cells. Treatment can be thought of as a game between a patient's doctor and their cancer. Accordingly, medical experts have begun using game theory to design better treatment protocols. If the doctor uses a single drug or strategy, the cancer may evolve resistance. We can think of this like rock-paper-scissors.

Perhaps a tumor is dominated by "rock" strategy cells. The doctor uses the "paper" strategy, wiping out the "rock" cells. But if the doctor uses only one strategy, then a previously tiny population of "scissors" cells may grow unchecked. The doctor should therefore cycle treatments. Next, she'll use "rock," and so on. Using game-theoretic modeling, medical teams can employ the dynamic strategies needed to battle a dynamic disease.

Price remained haunted by Hamilton's original question: What underpins the evolutionary origin of altruism? Hamilton had determined that altruism could emerge between related individuals, but Price wanted a more general solution. He started from scratch, eschewing as many assumptions as possible, and derived a simple equation describing natural selection. Now known as the Price equation, it's so general that it can describe any system changing in time. It expresses, in mathematical terms, the phrase "survival of the fittest." Given a trait of interest in a population—a mutation conferring better eyesight, say—the equation determines that the trait will become more common in successive generations if it increases fitness and less common otherwise. A second term in the equation represents the environmental effects on these changes. For example, systematic biases or mutations might affect the fitness of parents and their offspring. Or cultures that adopt altruistic norms might fare better.

These relationships seem obvious, but surprisingly, no one had yet captured this mathematically. Price's equation has provided a great deal of conceptual clarity to evolutionary dynamics. As biologist David Queller puts it, Price's equation revealed that "what we are seeking to explain—selection—is not a thing but rather a relationship, that between fitness and inherited values." It showed, for instance, that natural selection can operate without resource competition, long thought to be a necessary part of evolution thanks to Malthus's influence on Darwin. Evolution happens as long as gene variants spread through a population at different rates—there needn't be constraints on the population itself. Price shared this finding with the University College London statistics professor Cedric Smith, who immediately offered him an office at UCL. Price later wrote to his mother: "It was so simple that I felt sure someone must have discovered it before." The fact that an outsider to biology had derived this equation struck even Price as "quite a miracle."

The beauty of the equation lies in its generality. Price initially formulated it to describe the change in gene frequency over evolution but realized it was general enough to capture any change, even, as he wrote to a friend, selecting a radio station by turning a dial. It describes any kind of inheritance—not just genetic but epigenetic, behavioral, or symbolic. Price hoped to spark a new field: the science of selection. "A model that unifies all types of selection (chemical, sociological, genetical, and every other kind of selection)," he writes in a posthumously published paper, "may open the way to develop a general 'Mathematical Theory of Selection' analogous to communication theory."

Communication theory (better known as information theory) is the study of the fidelity of messages passed between receivers, and its formulation by Claude Shannon was foundational for the computing revolution. Once information had been defined, the field of information theory flourished. Likewise, Price hoped that defining selection would birth a new field. Selection, he writes, is involved in not only the fate of species but also that of languages, histories, economic growth, learning, and the practice of science itself:

Selection has been studied mainly in genetics, but of course there is much more to selection than just genetical selection. In psychology, for example, trial-and-error learning is simply learning by selection. In chemistry, selection operates in a recrystallization under equilibrium conditions, with impure and irregular crystals dissolving and pure, well-formed crystals growing. In paleontology and archaeology, selection especially favors stones, pottery, and teeth, and greatly increases the frequency of mandibles among the bones of hominid skeletons. In linguistics, selection unceasingly shapes and reshapes phonetics, grammar, and vocabulary. In history, we see political selection in the rise of Macedonia, Rome, and Muscovy. Similarly, economic selection in private enterprise systems causes the rise and fall of firms and products. And science itself is shaped in part by selection, with experimental tests and other criteria selecting among rival hypotheses.

Though a unifying theory of selection has not come to pass, scientists have used Price's equation to make headway in areas including genetic and epigenetic evolution, social evolution, language change, drug interactions, niche construction, and more. Price's notion that selection acts on ideas presaged Richard Dawkins's memes: ideas as entities that compete for mental and cultural representation. The philosopher Karl Popper also likens the acquisition and generation of knowledge to biological selection, describing the process as "evolutionary epistemology." New ideas are like genetic mutations. The brain brings forward several proposals, and some selective procedure eliminates whichever possibilities are unacceptable to it.

However, Price failed at his goal of discovering the biological basis for pure selflessness. Hamilton's rule, describing how a gene can give rise to altruistic behaviors, turned out to be a special case of Price's more general formula. The Price equation indicates that altruistic behaviors will be favored over time so long as they sufficiently benefit the population overall. Altruism is just as inevitable a consequence of natural selection as selfishness. But spite can also be favored: an individual might harm another, even at personal cost, if they are less related than the population average. All that matters is whether these traits increase their bearer's fitness; it's a fundamentally selfish selflessness. Inherent in our conception of altruism is a notion of sacrifice, of loss. It's not *genuine* altruism if something is gained. Perhaps a more hopeful framing, however, is to marvel that the structure of life is such that, by helping others, one may also help oneself. No creature is an island; no one is divisible from the greater whole.

Price personified the danger of taking our mathematical models too seriously. He was profoundly disturbed by his equation and its suggestion that altruism is subsidiary to selfishness. The poet John Keats lamented that Isaac Newton had "destroyed the poetry of the rainbow by reducing it to a prism." Similarly, Price despaired that his equation had stolen all the goodness from altruism. His discovery left him by turns dazzled and horrified. The formerly militant atheist became haunted by revelatory visions of Jesus, likely brought on by his poor health. He became obsessed with serving others. In an ecstatic revelation that would precipitate the end of his life, George Price heard Jesus whisper: "Give to everyone who asks of you."

The world became a great test of his faith, and he removed every guardrail from his life. He failed to renew his UK visa or grants and eschewed eating as much as possible. He quit his thyroid medication, believing that this made him more receptive to God's counsel, taking literally Jesus's Sermon on the Mount admonition to "take no thought for the morrow." He gave away everything he could to disprove his own equation, or at least empirically strain it. He introduced himself to people living on the streets of London: "My name is George. Is there any way I can help you?" His worried colleagues at UCL begged him to come stay with them. Maynard Smith wrote: "I have less faith than you do that the Lord will provide. *Please* let me know at once if I can help."

The Lord did not provide. Within months, Price lost his apartment. Having given away everything, he worked as a night janitor, finishing one final paper with Hamilton in his spare time. But his call to

Christian service outshone his worldly duties. He despaired that even as an altruist he was a failure—he hadn't measurably helped *anyone* yet had still managed to ruin himself. He was living in a squatter's dorm with artists, alcoholics, and revolutionaries, in love with a print-maker half his age who refused his marriage proposals. Hopeless of resolving his ethical quandary, his teeth rotting, his nails filthy, and his waxy skin jaundiced, Price took his own life in the squatter's dorm by snipping his carotid artery with tailor's scissors.

Police called in Hamilton to identify the body. Price's funeral was sparsely attended by four squatters from his community and his for-mer collaborators, Hamilton and Maynard Smith: two of the most celebrated biologists of the twentieth century. Price's housemates con-fessed that they had no idea he'd been such an accomplished scientist, and Price's grave remained unmarked for decades. He had published so sparingly that his contributions were often attributed to others. May-nard Smith admits, in a collection of essays, that he might not have written his career-defining papers if he hadn't seen Price's unpublished manuscript on the evolution of limited combat. "Unfortunately," he writes, "Dr. Price is better at having ideas than at publishing them. The best I can do therefore is to acknowledge that if there is anything in the idea, the credit should go to Dr. Price and not to me." Maynard Smith's self-effacement was exaggerated—his work was foundational for evolutionary game theory. Nevertheless, it's clear that Price was working at the forefront of this new field. His research and ensuing moral crisis read like a metaphor for the twentieth century's reconcili-ation of biology and ethics. Hamilton describes his life as "a completed work of art."

Hamilton believed that Price's obscurity was intentional: Price's burgeoning Christian humility had translated to an excessive reserve in his papers. Chastened by faith, he feared overclaiming, resulting in a particular economy of prose. He wanted his work to reveal its depth only to those willing to put in careful contemplation. He'd rely on others to expound on the tidily encrypted ramifications of his equations—much as, Hamilton speculates, Jesus's apostles had clarified his spare gospel. "In this process, I believe I was chosen to be his first initiate," Hamilton writes. Hamilton died twenty-five years later, still adamant that Price's equation had yet to unveil the full extent of its beauty to the world.

This remains true today: Price's quest to build a general science of selection stands unrealized. The Price equation is the fundamental theorem of evolution and can be used to derive other basic theorems of evolution. The equation's generality lends it its power. This same generality invites misuse from researchers unfamiliar with its nuances. Some have dismissed it as a tautology, nothing more than a statistical identity. Others have praised its depth. The biophysicist Steven Frank has argued that the Price equation shares a deep consonance with several distant fields. The fundamental equations of thermodynamics, physical mechanics, and Bayesian updating, for instance, all involve the maximization of a quantity (e.g., entropy, momentum, and information) in a manner analogous to the role of fitness in Price's work. Frank contends that Price's equation offers a "powerful constraint on the geometry of change" and "hints at a deeper common mathematical structure" unifying these fields. There may be a deep formal relationship between natural selection and information theory, entropy, probability, and other fundamental physics, all captured in the architecture of a deceptively simple equation.

Scientists discovered that the counterintuitive process of evolution—harnessing chance to accrue order—was best expressed in the language first developed by gamblers. Darwin's idea seemed impossible at first blush. It seemed absurd to speculate that an organ as exquisite as the eye could arise from chance. But life works by way of immensity. Over the vast timescales of evolution, different species accrued their particular genius. Natural selection is a process that generates high improbability. The "impossible" sex ratios Arbuthnot calculated turned out to be the result of neither chance nor design but a game of selection billions of years old.

While many biological entities don't have brains per se, here players needn't have minds, only desires. Every living entity has wants—they all need energy and the conditions required to maintain themselves. Given two living entities whose desires somehow interact—one eats bacteria, say, and the other eats amoebas that eat the same bacteria—a game arises. Games precede brains—in fact, they are foundational to the emergence of brains, as species spun up ever more elaborate countermeasures to outdo one another over evolutionary history. We might think of games as shaping what it means to be a self: we're defined in opposition to our playmates. The self discovers its borders in relation to others. The past four hundred years of scientific discovery have radically unsettled humankind's conception of its cosmic position. Astronomy unseated us from the center of the universe, geology displaced us from the center of time, and biology ousted us from the center of life. We became, instead, mere effects of immense and indifferent physical processes. We do not play the game of life; the game of life is what plays us into being.

Much as economists had despaired that their field might never become a rigorous science, biologists thought, for a time, that the study of life was doomed to remain a collection of facts. Game theory gave biology, as it had economics, a theoretical backbone. But game theory is an imperfect framework for evolution for the same reason that it's an odd fit for economics: it's static; it describes equilibrium points. Evolution is the process of change. Creation is, as Fisher puts it, "still in progress, in the midst of its incredible duration." Even so, game theory can model life's destinations: evolution "ends" at a Nash equilibrium. It details the balance of nature, long present in folk wisdom. Importantly, the Price equation also demonstrates that evolution doesn't require resource competition, and Darwin's Malthusian undertones can be discounted. In recent decades, evolutionary scientists have increasingly focused on the many ways cooperation is also a fundamental principle of life.

Statistics and game theory transformed evolutionary biology from a historical, descriptive science into a predictive one. Biologists once cataloged the endless variations of life. Now they can posit explanations for why some of these strategies arose and forecast other potential equilibria. Unlike physics and chemistry, with their passive laws, biology requires its students to consider purpose. These ideas might even help us more clearly define life and draw a line between the living and nonliving worlds. Games helped systematize a complex field, and game theory was expanded in return. Instead of being bound by strict assumptions of rationality, necessitating infinite foresight and computational resources, a player can be something much simpler: a fir tree, an amoeba, an Arabian warbler. Players don't choose their traits or strategies—they inherit them. Things without intelligence can play games. Perhaps intelligence itself can be bootstrapped from playing games.

Nous ex Machina

In a riddle whose answer is chess, what is the only
word that must not be used?

JORGE LUIS BORGES

laude Shannon, the founder of information theory, argues in a
1950 paper that games offer an ideal environment for developing
machine intelligence. Up until that point in time, researchers had
mainly used computers to crunch numbers in massive calculations.
Instead, Shannon argues, computers could one day process "chess
positions, circuits, mathematical expressions, words." Games aren't
practical, he grants, but they could be a training ground for tackling
other problems, like translating texts into different languages, making
strategic military decisions, and composing music—in other words, for
simulating human intelligence. In the paper, he presents an early
chess-playing program as an attempt to approximate thought.

Shannon saw chess as the ideal starting point for this project. Chess
is a well-defined realm. Compared with real-world problems, chess is
easier to manage, and progress within the game is more readily measured. Games are discrete, and play appears to require the hallmarks of
intelligence: strategy, planning, and logic. Games are a way to render

intelligence legible; they give researchers a vocabulary to articulate the invisible powers of the mind. Shannon wasn't the first person to have this realization: history's earliest programmer, Ada Lovelace, wondered whether a game like solitaire "admits of being put into a mathematical Formula, and solved." She was the earliest scholar to grasp that Babbage's computing machine might do more than simple numerical analysis; it might analyze and even create games, texts, and music.

A game is like a branching web of possibility. At the start, players are faced with a tree of all mathematically legal plays. Each turn prunes the space of next possible moves. In the limited realm of gameplay, a player's intelligence might be boiled down to two fundamental components: the efficient search of all possible futures and the selection of the path most likely to yield a win. To navigate those possible futures, Shannon turned to game theory. Von Neumann argued that rational players make choices they believe will maximize their future payoffs. Similarly, Shannon proposed building a chess program that used minimax estimations to steer its choices toward a win. How would the engine determine which move was most promising? Chess doesn't have a score per se, but a human player can look at a board and roughly estimate whether they are winning or losing and reason whether a given move will help or hinder them. Shannon equipped his program with a hard-coded evaluation function to choose moves, replacing human judgment.

The evaluation function became a key component of early gameplaying systems. It required researchers to introspect about how they themselves navigated play, mapping the anatomy of their inner reasoning and positional judgment. Researchers strung together complex mathematical formulas summing up factors deemed important by chess experts: the number of pieces remaining, their positions on the board, the king's vulnerability, isolated pawns, and so on, each weighed by its relative importance. A bishop might be worth three pawns, for instance, but half as much as a queen. Shannon's proposed program would calculate this value for all legal next moves, then repeat this evaluation for the opposing player's response. It could be assumed that the opposing player would want to minimize the first player's score, and the program would predict their move accordingly. The program would continue to estimate outcomes for all possible sequential responses over the next several turns, then pick the move most likely to lead to a win, assuming that both players would want to minimize each other's scores.

Given the computing technology of his time, Shannon estimated that fully elaborating all possible chess games would take much longer to calculate than the known age of the universe. But his goal wasn't to make a program that could play a perfect chess game, just a skillful one. In his paper on the endeavor, Shannon jokes that if players truly had the idealized foresight assumed by game theory, chess would hardly look like a game:

A game between two such mental giants, Mr. A and Mr. B, would proceed as follows. They sit down at the chessboard, draw the colors, and then survey the pieces for a moment. Then, either:

(1) Mr. A says, "I resign," or

(2) Mr. B says, "I resign," or

(3) Mr. A says, "I offer a draw," and Mr. B replies, "I accept."

In practice, a program could still do well by calculating just a handful of future steps. After all, even chess masters admitted to looking only a few steps ahead. As a proof of principle, Shannon outlined the design of a simple program that itemized all legal next moves. It then chose randomly among those options. He played chess against his algorithm, calculating its moves by hand. He had won within five turns, dutifully reporting that "the level of play with such a strategy is unbelievably bad." Yet he recognized that computers would improve and eventually provide faster, error-free, indefatigable calculations for analyzing game positions. "These must be balanced against the flexibility, imagination and inductive and learning capacities of the human mind," he writes.

Across the Atlantic, Alan Turing had been contemplating the same problem. By the late 1940s, he and his colleague David Champernowne had designed a program called Turochamp, "based on an introspective analysis of my thought processes while playing." Similar to Shannon's design, Turochamp calculated all potential moves and responses. It then assigned point values to each based on the mobility and safety of important pieces, and used minimax to select the move that yielded the highest expected value. It wasn't feasible to run the program on the computers of the time, however, so they calculated their algorithm by hand. In 1952, they pitted the Turochamp program, tallied on paper, against their colleague Alick Glennie. Turochamp fared better than Shannon's program, losing after twenty-nine moves, each taking half an hour to compute.

Games subverted a significant challenge in the study of intelligence: the need to strictly define it. Turing famously proposed the Imitation Game, better known as the Turing test, to gauge a program's intelligence. Here, a human judge communicates with a language program or another human by typed text. In Turing's view, a computer program that can convince the judge it is human should be deemed intelligent. But this is vulnerable to subjective interpretation: judges might make different assessments, colored by their own experiences and assumptions about their interlocutor. Games like chess, on the other hand, are unambiguous. They have clear outcomes, unstained by interpretation. You win or you lose. By imbuing a computer program with a goal—winning a game—researchers could track clear progress in their pursuit of defining intelligence. This was hugely attractive in the precarious world of academic research, which, by definition, involves plumbing the unknown while simultaneously providing clear benchmarks to satisfy funding bodies and investors. Competing research teams raced to build the first program that bested humans in popular board games. Using games as a substrate for artificial intelligence gamified the realm of research itself. However, our understanding of what intelligence entails has been impoverished by this focus on games.

AI continued to be heavily influenced by game theory. For decades, researchers relied on minimax to steer their game engines' choices. But this reduces intelligence to an economic calculation. A game-theoretic agent's goal is to maximize its payoff. Similarly, a prominent definition of intelligence put forward by AI researcher John McCarthy has it as "the computational part of the ability to achieve goals in the world," echoing the game-theoretic agent's purposefulness. Researchers became fixated on building systems that could achieve clearly specified goals. The cost-benefit analyses that shaped economic theory came, somewhat regrettably, to stand as a proxy for thought, confounding intelligence with the single-focused pursuit of measurable material gain. Other forms of genius—physical, emotional, linguistic, musical—are difficult to measure or model, and so were ignored. Self-awareness and empathy were similarly omitted. Researchers are now waking up to the dangers inherent in making systems that single-mindedly focus on goals without the ability to reason about possible side effects. A doctor hoping to treat a patient's cancer mustn't kill the patient in the process.

Despite decades of work on game-playing programs, by the 1980s, computers had only bested humans in the simplest of contests, like tic-tac-toe. But the pursuit led to major advances in computational techniques that remain standards of the field to this day. For example,

Arthur Samuel started his checkers program before standard programming languages existed, so he invented a new language as he needed it. Game researchers also developed tree search methods that could project out all possible moves of a game—techniques that proved handy for other complex data analyses. Because games require combing through a massive search space, their study spawned foundational ideas about how to make computing more efficient.

Humans still reigned supreme in most games, however. After decades of development, Samuel's self-playing checkers program was still no match for top human players. A checkers expert who'd played the engine advised Samuel to add a library of first moves and endgame problems. Samuel declined: "I cannot do that. That reduces the game of checkers to simple table lookup. I did not teach the computer how to play checkers; I taught it how to learn to play checkers." This attitude had become rare among AI researchers. The problem with gamifying the pursuit of understanding intelligence was that many researchers became more concerned with winning games, even by brute force, than understanding intelligence. Samuel didn't want to build a system that could win at checkers; he wanted to build a learning system whose principles were relevant to more important problems. "It seems reasonable to assume," he writes in a 1960 paper summarizing the state of machine learning, "that these newer techniques will be applied to real-life situations with increasing frequency, and that the effort devoted to games or other toy problems will decrease."

Computer game demonstrations were reduced to progress theater: publicity stunts used to drum up attention or investment for their patron corporations and universities. These systems were tailored only to the games they'd been created to play, an intricate tangle of dependencies that an unexpected input could easily shatter. Researchers like

McCarthy abandoned computer chess research for this reason. "Unfortunately, the competitive and commercial aspects of making computers play chess have taken precedence over using chess as a scientific domain," he writes. "It is as if the geneticists after 1910 had organized fruit fly races and concentrated their efforts on breeding fruit flies that could win these races."

While Samuel refused to use a lookup table in his checkers program, other researchers would fully solve the game by this method. Checkers is simple enough to be entirely determined by computers, rendering it a pure mathematical object. This still took decades of work. In the late 1980s, the computer scientist Jonathan Schaeffer started his research lab at the University of Alberta and decided to focus on checkers. If intelligence can be distilled to the process of making sensible choices in a thicket of possibilities, checkers is much more tractable than chess. In chess, there are about thirty-five possible moves in a turn; in checkers, about three. To enumerate all possible moves over three turns in chess, one must search over forty-three thousand possibilities. For checkers, this is more like twenty-seven. A checkerboard "only" has 10^{20} possible positions, compared with 10^{44} possible chess configurations.

Schaeffer's program, Chinook, navigated the tree of possible choices using minimax and judged best positions using an evaluation function informed by chess experts. Unlike Samuel's program, Chinook couldn't learn on its own. Its actions were preprogrammed, strictly determined by its evaluation function and database of early- and endgame moves. In checkers, the probability of winning strongly correlates with simple metrics, like how many pieces remain for each player, how many pieces are kinged, and so on. To make searching through the tree of all possible moves more efficient, Schaeffer used a now ubiquitous pruning method called alpha-beta tree search, discovered independently by several groups working in games. This prevented the program from exploring moves that led to outcomes worse than the outcomes of moves it had already searched. For example, by calculating ahead some number of turns, a program might discover that a move could result in losing a piece. This move would not be further explored, in favor of exploring better possibilities more deeply. However, this results in programs that have a conservative and materialistic play style. Because these programs stop searching branches that yield any loss, they can't strike on strategies in which a temporary sacrifice leads to a greater gain down the line.

Chinook was not lacking foresight, however: it could search dozens of turns ahead. It won the 1989 Computer Olympiad in London. There, Schaeffer met a checkers expert, Herschel Smith, who had followed Chinook's games with great interest. Smith had played Samuel's program decades earlier and was enthralled by Schaeffer's progress. Schaeffer, Smith realized, knew almost nothing about advanced checkers play. This made Chinook's achievement all the more impressive. If Schaeffer really wanted to prove his program, Smith advised, Chinook would have to play against Marion Tinsley. Tinsley, by then in his sixties, is widely considered to be the greatest checkers player of all time. Over the forty years he played competitively—in thousands of tournaments and demonstrations—he lost fewer games than he could count on one hand, and never twice to the same person. He claimed to have spent about ten thousand hours studying checkers while in graduate school for math (largely, he admitted, in lieu of doing math). He had an uncanny memory and could recall every move in every game he'd ever played.

Tinsley was a devout Christian. He dressed tidily in somber suits and dark-rimmed glasses, though his favorite tie pin spelled out "JESUS" in multicolored stones. He lived with his mother until her death, never marrying. "I haven't seen a checker marriage that worked out. It is a very rare woman who can be married to a real student of

208 Playing with Reality

checkers," Tinsley solemnly reported. The game was his one earthly love. "Chess is like looking out over a vast open ocean," he said. "Checkers is like looking into a bottomless well."

Tinsley agreed to play Chinook in several exhibition matches in the early 1990s. He remained serenely confident. "I have a better programmer than Chinook," he told a reporter. "His was Jonathan, mine was the Lord." Later, he confessed: "I don't want to let my programmer down, and I'm sure I won't." Though Tinsley won these matches, it was clear that Chinook was much stronger than programs that had come before. Most of their games ended in draws. Chinook only lost one game and came close to beating Tinsley in another.

After Schaeffer moved a piece for Chinook early in one game, Tinsley looked up in surprise and said, "You're going to regret that!" Schaeffer recalled thinking, "What's this guy talking about? We're in no trouble." Chinook had been projecting that the game would end in a draw, and the move had not altered its estimation. Forty-six turns later, calculating nineteen steps ahead, Chinook began to forecast a loss. "Can Tinsley possibly be looking sixty-five plies ahead?" Schaeffer wondered, then dismissed the idea. Yet Chinook went on to lose the game. After the match, Schaeffer analyzed the game and discovered that, assuming both players played perfectly, every outcome after the move Tinsley mentioned would result in a loss for the player who made it. Tinsley's depth of understanding was beyond machine reckoning. After fervently analyzing Tinsley's games, Schaeffer was shocked at how closely he played to verifiable perfection.

Chinook played Tinsley in several more matches, always close in skill but never quite beating him. Before their final rematch in 1994, Tinsley had a troubling premonition. He reported the vision to Schaeffer the following morning: "I had a dream last night [. . .]. In it, God told me that he loves you, too." On the second day of the match,

Tinsley resigned due to stomach pain, which was soon discovered to be cancer. He died the following year.

Beating Tinsley had been Schaeffer's sole ambition. Pitting Chinook against any other checkers champion seemed pointless. Bereft of purpose, Schaeffer shifted his research focus. Chinook would no longer play against humans in tournaments. Chinook would *solve* checkers. At the start of a game of checkers, with twenty-four pieces on the board, there are about 10^{20} possible positions. By the endgame, with only ten pieces left, about thirty-nine trillion possible positions remain. Schaeffer's team first simulated and cataloged the entire gamut of these possible endgames. Chinook then searched for paths that connected all possible checkers openings with the endgame library—a feat that required over a decade of brute-force computing.

Chinook's calculations finally ground to a halt in 2007. "It's done," Schaeffer reported to his daughter. "Checkers has been solved." She recalled how "he gazed around as though this were the most momentous occasion in history and would have a picture of this moment right beside the one of the first men on the moon." Schaeffer and his team had found that a game of checkers will always end in a draw if both players play perfectly. Though impressive, Schaeffer's brute-force solution left science none the wiser about the nature of intelligence. It rendered checkers into a cascade of possibilities, defined by the game's rules and their logical consequences.

Unlike in checkers, the variations of a chess game are too huge to fully reckon. To tackle chess required the invention of vastly more powerful computer hardware. For those unable to peer through the Iron Curtain during the Cold War, even impractical intellectual pursuits had come to stand as bellwethers for the superpower's technological supremacy. The Soviets had dominated world chess titles for decades, save in 1972, when the American Bobby Fischer won against

Boris Spassky. The game was so politically fraught that Henry Kiss-
inger, then Nixon's national security adviser, had personally called
Fischer before the match to encourage him: "America wants you to go
over there and beat the Russians." The chess grandmaster Garry Kasparov later played up the political symbolism of this win, claiming it
"was treated by people on both sides of the Atlantic as a crushing moment in the midst of the Cold War."

For a time, it seemed that Soviet researchers would also dominate
in computer chess. The race to build AI was not purely for posturing,
of course: military and political leaders hoped to shunt morally charged
decisions to machines. Former world chess champion Mikhail Botvinnik had spent decades perfecting his chess program, Pioneer, a program he hoped would one day be intelligent enough to manage the
Soviet economy better than imperfect human planners or the free
market's inequitable myopia. Faced with the limited computing power
of the era, Botvinnik used his expertise to strictly limit the moves that
Pioneer considered at each turn of its search. By the early 1980s,
rumors were circulating about an impressive new American contender, Belle, created by Bell Labs researcher Ken Thompson. Botvinnik invited Thompson to Moscow to demonstrate his program, and
Thompson gladly accepted.

The demonstration never happened. Thompson learned upon his
arrival in Moscow that Belle had been seized by US officials and never
made it out of the country. Belle wasn't just a program; it was a computer system. Instead of restricting the program's search capabilities,
as Botvinnik had, Thompson had designed more powerful hardware.
Already famous for inventing the operating system UNIX, Thompson
was a standard fixture at computer game tournaments. Schaeffer recalled Thompson taking him under his wing at one of his earliest
matches, offering sage advice with unmetered generosity while wearing

a T-shirt emblazoned with the "simple picture of a large, fat cat." Upon his return home from the USSR, Thompson faced arrest on suspicion of exporting advanced technology to the Soviets. A reporter covering the story asked Thompson if Belle had any military use. Thompson pondered for a moment before replying: "Maybe if you threw it out of a plane, it could kill somebody."

In 1983, Belle became the first computer system to earn the title of US national master, spurring other researchers to hasten their efforts. Feng-hsiung Hsu studied Belle's architecture as a graduate student at Carnegie Mellon and engineered a chip twenty times faster and a thousand times cheaper. An array of these chips could search millions of moves in parallel. Hsu and his colleagues' work—performed on a shoestring academic budget—was so impressive that IBM snatched up the entire team in 1989 to continue the project.

Their chess-playing supercomputer would come to be known as Deep Blue, named in honor of the Deep Thought computer in Douglas Adams's *The Hitchhiker's Guide to the Galaxy.* Deep Blue's program was not qualitatively different from what had come before. Its main distinction was its massive computing power. Deep Blue chose moves based on the same minimax principle proposed by Shannon in 1950. It couldn't learn independently, working more through brute force than elegance. It explored up to two hundred million possible positions per second. Once a behemoth supercomputer housed in a pair of two-meter-tall cabinets, Deep Blue today could be run on a smartphone.

What constitutes a promising move is much trickier to assess in chess than in checkers. It involves the identities of remaining pieces, their legal moves and positional strength, the myriad potential interactions of the pieces at hand, and countless other factors. Deep Blue's final evaluation function had over eight thousand components, a spiderweb of conditional dependencies. The researchers tinkered with the weights of every variable: How much should they value a knight, and how does this depend on its rank? How does this relate to the other player's remaining pieces and their relative positions?

The program could "see" only as far as its programmers could imagine. All possible chess games have not been played, nor will they ever be. Chess is a historical and cultural enterprise: millions of players have studied the same famous games, emulating prevailing styles and gambits. Deep Blue and other chess programs were subject to the same education. Unlike people, who can usually figure out what to do when they encounter something new, these programs' decisions were rigidly prescribed and therefore prone to failure when facing unusual moves or rare board configurations. Human players took advantage of this by employing a style that came to be known as "anti-computer chess." For example, players traditionally vie to control the center of the board, but a player using anti-computer tactics might move to the board's edge to throw off their programmed opponent. This strategy can also work for human players: early chess games have been so relentlessly optimized that players tend to stick to standard openings. World champion Magnus Carlsen is known to open with nonstandard moves to force his opponents outside these well-practiced beginning states. IBM paid chess experts to play against Deep Blue in unconventional ways to refine its evaluation function and surface any easily exploitable bugs.

In 1996, after a decade of development and training, Deep Blue was pitted in a match against then reigning world chess champion Garry Kasparov, still considered one of the greatest players in history. Though Deep Blue lost the match, it was the first computer program to win a game against a human champion, which it did thanks to a masterfully brash gambit. Deep Blue captured one of Kasparov's pawns while temporarily leaving its king exposed. Because it could calculate fifteen moves ahead, Deep Blue had determined that it would be able to defend this vulnerability by a hairbreadth. It was a move, commentators argued, that no human would have risked. "In certain kinds of positions," Kasparov remarked, "it sees so deeply that it plays like God." Computer scientist Douglas Hofstadter was stunned by the win: "My God, I used to think chess required thought. Now I realize it doesn't. It doesn't mean Kasparov isn't a deep thinker, just that you can bypass deep thinking in playing chess."

Deep Blue won its 1997 rematch against Kasparov. In the second game, it made a move so beautiful that Kasparov was convinced a computer couldn't have made it. He was sure that one of the chess experts consulting for IBM had intervened: "It reminds me of the famous goal that Maradona scored against England in '86," Kasparov alleged. "He said it was the hand of God." Argentine soccer player Diego Maradona had, it was later established, cheated, guiding the ball into the net with his hand rather than his head. Cheating wasn't unheard of in computer game tournaments. A program's human operator usually moved pieces for their program. Some had been known to substitute their own move for one of the computer's to save face from a particularly embarrassing program error. Kasparov demanded to see Deep Blue's logs. The IBM team refused, claiming it would be too revealing of Deep Blue's "inner thoughts." The program printed its reasoning alongside each of its move choices. Kasparov grew even more convinced of foul play, though the team later shared its printed logs with a neutral arbiter. Kasparov was, to be fair, at a disadvantage. Deep Blue had studied Kasparov's entire career. Every move he'd ever made had been salted away in its prodigious memory. Kasparov was not allowed to see any of Deep Blue's games and so had no way to prepare for his match.

For the first time, a human grand champion had lost a chess match to a computer program. Chess commentator Miguel Illescas complained that Kasparov had played as though afraid. Kasparov replied:

I'm not afraid to admit I am afraid, and I'm not afraid to say why I am afraid. It goes beyond any chess computer in the world. I'm a human being, you know. [. . .] When I see something that is well beyond my understanding, I'm scared.

The IBM team was devastated by the allegations of cheating and the vilification by the press. The audience booed Deep Blue's victories—humanity wanted Kasparov to win. Hsu found it baffling: "The match was never really 'man versus machine,' but rather 'man as a performer versus man as a toolmaker.'" While the public was openly hostile to Deep Blue, chess experts were electrified and praised the program for illustrating new endgame tactics.

Though the machine was an inspired feat of engineering and a computing breakthrough for its time, it's unclear what Deep Blue—a narrowly focused, baroque engine suited to playing chess and nothing else—ultimately contributed to our understanding of intelligence. Like Chinook, Deep Blue demonstrated that certain aspects of planning can be mimicked through brute computational force, navigating the many paths of a massive tree search toward a favorable outcome. After the match, a reporter asked Joe Hoane, one of Deep Blue's programmers, whether the team had tried to emulate human thought. Hoane replied: "No effort was devoted to [that]. It is not an artificial intelligence project in any way. [. . .] We play chess through sheer speed of calculation and we just shift through the possibilities and we just pick one line."

Deep Blue could only reflect the knowledge programmed into its evaluation function; it couldn't surpass human understanding. Like Chinook, it could only spit out game moves. Murray Campbell, another member of the Deep Blue team, looks back on the effort, decades later, with some regret:

In the past few years, I've started thinking differently about games like Go and chess and certainly checkers. As hard as they are for people to play, they are, in hindsight, not that interesting for AI. [. . .] I guess the question is, "Are two-player zero-sum games of perfect information the best domains for continued AI research?" I think not. If you ask me what code that we took from Deep Blue and applied to other problems, there wasn't anything.

If nothing else, games continued to be an unparalleled means for engineers to drum up attention and funding. In the months following the match, IBM's stock price rose by about 40 percent. Deep Blue was dismantled after its match with Kasparov, but it inspired a slew of programs that came to outrank it. Stockfish, an open-source community project with scores of contributors, replaced Deep Blue as the reigning world computer chess champion. The top-ranked chess program for years, it stood as a testament to the power of the open-source community. Volunteers donated spare CPU time to train the model in parallel, bypassing the need for supercomputers. It was similar in architecture to Deep Blue: a complex organ with dozens of separate working parts and an evaluation function conditioned on specific board configurations. Like Deep Blue, it was good at only one thing: chess. Advancing computer gameplay did not have the intended side effect of understanding intelligence, though games and computer science enriched each other in a virtuous cycle. These advances would later feed back to the development of more intelligent computer programs in unpredictable ways.

In 1962, three MIT graduate students built a tool to demonstrate the capabilities of a powerful new computer. The result was *Spacewar!*, the world's first video game. In it, players control a spaceship engaged in a dogfight while trapped in the gravity well of a star.

Spacewar! became so popular that Stanford computer labs had to ban its use during business hours so that students would still get research done. Nolan Bushnell played it while still an engineering student at the University of Utah and was instantly smitten. He'd been working as a self-described carny to fund his way through college, hawking mechanical games at Lagoon, a local amusement park, and immediately saw *Spacewar!*'s commercial potential. "I knew that if I could put a coin slot on that game that it would earn some money in one of my arcades," he said. Bushnell had little hope of capitalizing on it, though: computers at the time cost hundreds of thousands of dollars. They were, however, getting cheaper.

In 1972, a few years after graduating from college, Bushnell founded Atari Games, named after the term used by Japanese Go players that is analogous to chess's "check." Atari's first game, *Computer Space*, was inspired by *Spacewar!*, built with custom hardware housed in a glittery midnight-blue cabinet. Bars bought a few thousand cabinets to cater to drunk patrons looking for a futuristic alternative to darts. But the complicated game never took off. Bushnell and colleagues sought to make their next game as simple as possible, and *Pong* was born. Bushnell charged two of his employees, Steve Jobs and Steve Wozniak, with the design of a one-player version of *Pong*, the now classic *Breakout.*

On the side, the Steves were developing a project they'd hoped would wow their boss: a home computer system they'd crafted using spare microprocessors. When they pitched the Apple I to Bushnell, he was too distracted to be interested: Atari was rolling out home consoles, and Bushnell was negotiating a buyout by Warner Communications. Nor, some months later, would he be interested in investing in their splinter organization, Apple—Jobs had offered him a stake of one-third of the company for $50,000. (Today, this would be worth about $1 trillion.) In 1977, finally relieved of his obligations to Atari,

Bushnell was free to pursue, with single-minded focus, his true dream, one that had come to him in a vision while still a young carnival worker: Chuck E. Cheese's Pizza Time Theatre. He believed that Chuck E. Cheese—his grandest vision yet, a ministry of chaos—would be the vehicle to take video games mainstream.

Previously, arcade games had been the purview of bars and pool halls, orbited by surly teens and dishonored, in Bushnell's estimation, by their proximity to mechanical peep-show machines. Given their earliest habitat, Atari's games were purposefully simple, largely sans text, so that "any drunk in any bar" could play them. They were also, therefore, ideally suited for children, and this was the audience he most wanted to attract. With the sensory exuberance of Chuck E. Cheese, he sought to satisfy what he believed to be a primal need of humans: "Whether you were talking about the summer solstice with primitive man to the circuses in Rome, there was always an entertainment element." Bushnell would be betrayed by his codeveloper, Robert Brock, who, having decamped with a superior animatronics engineer, reneged on their agreement and used what he'd learned about running Pizza Time Theatre to establish the rivalrous ShowBiz Pizza Place in 1979. The video game crash of 1983, caused by a glut of new companies and mediocre games flooding the market, left Chuck E. Cheese's Pizza Time Theatre racked with debt, its stock price 90 percent off the highs. By 1985, ShowBiz Pizza had finalized a corporate takeover of Bushnell's beleaguered dream, though his video game empire would leave a lasting mark on computing and AI.

Bushnell's intuition had been correct: children would become the prime driver of the video game market. However, the vector of mass adoption was not Chuck E. Cheese but home computers and consoles. While Bushnell was busy battling corporate intrigue, the 1970s saw the birth of personal computers—thanks in large part to the Steves— and the popularization of more user-friendly programming languages, like Basic. In 1978, the book *Basic Computer Games* became an unlikely hit and introduced a young generation to the principles of programming. It contained recipes for one hundred simple games and instructions on how to run them. Children became fluent in code before they reached college. A generation of kids who had cut their programming teeth on games went on to found tech companies. One of Bill Gates's earliest programs was a game that played tic-tac-toe. Young Sundar Pichai developed a chess engine; a twelve-year-old Elon Musk wrote a *Space Invaders* knockoff called *Blastar*. Before becoming the CEO of Salesforce, Marc Benioff sold his game creations to Atari, including *Crypt of the Undead* and *King Arthur's Heir*. The teenage creator of Ethereum, Vitalik Buterin, learned to code by building games. Video games were the emissaries of computers, enticing millions of young players into engineering.

Consumer demand for video games also incentivized the development of the graphics processing unit, or GPU. High-quality video displays were needed to render increasingly sophisticated computer game graphics. In the early 2010s, researchers realized that GPUs are hundreds of times more efficient for training neural networks than CPUs, given that they're suited for processing large data arrays in parallel. Engineers used this happy accident to train AlexNet, the 2012 image-labeling program that marked an inflection point in computer vision. It's now believed that early AI research was held back more by technical limitations than by conceptual ones. Today, companies like OpenAI operate on the assumption that intelligence will emerge simply by scaling learning programs with enough processing power.

Most definitions of intelligence include a qualification of generality: an intelligent agent should perform well on a broad range of tasks. Deep Blue could best a world chess champion but was utterly incapable of playing much simpler games, like checkers and tic-tac-toe. If engineers wanted to re-create intelligence, they needed to build more general programs capable of learning, rather than simply following preprogrammed rules. Video games offered a more diverse curriculum. In 2012, computer scientist Michael Bowling and his colleagues released a standardized collection of Atari games as a playground for training AI programs. Atari games are simple, given the limited processing power of computers in their era. But there are dozens of them, each requiring different skill sets. *Breakout* requires fast reflexes. *Asteroids* requires the player to build an intuitive model of the vehicle's physics to successfully steer around asteroids. *Pitfall* requires the player to explore the world before they can score any points. Handily, video games have been explicitly designed to teach people how to play them through simply playing them.

In 2013, a then obscure London start-up called DeepMind used Bowling's platform to develop a reinforcement learning agent that could learn to play dozens of different Atari games at or above human-level performance. Researchers tasked the program with increasing its score, then fed it raw game images and left it to train for thousands of years of sped-up gameplay. The program learned to choose the right actions to control its avatar without being explicitly told its goals—for example, "You are the submarine, which looks like a lumpy yellow rectangle, and you should hit those enemy ships, which look like lumpy gray rectangles." This sparked a resurgence of learning-based systems. Companies would go on to tackle increasingly complex video games, such as *Dota* and *StarCraft II*, whose realism better approximates real-world use cases.

Games remain an impressive way to evaluate a program's capabilities. In the early 1980s, graduate students Shafi Goldwasser and Silvio Micali invented a powerful new form of mathematical proof by imagining how they might play a secure poker game over the phone. The resulting technique, interactive proof systems, consists of two subparts, a prover and a verifier. These exchange messages until the verifier is "convinced" that an answer is correct. Interactive proofs became a foundational technique in computer science and cryptography. They're also the metaphoric heart of how we hope to verify intelligence through games. Traditional math proofs are inevitable, each step an irresistible conclusion unfolding from the last. Interactive proofs, on the other hand, are probabilistically true. By testing again and again, the verifier becomes increasingly convinced of the prover's claim. The Turing test, for instance, can be mathematically formalized in this framework. Games take the same form: human experts repeatedly test a game-playing computer program to convince themselves of its capabilities. There is no single measure of intelligence, or its mechanical approximation. Its evaluation will be, by nature, interactive: the iterated probing of a system.

Game theory–based minimax programs were good enough to beat top human players at several board games, but few would call these programs meaningfully intelligent. Games did, however, advance the state of the art for computer chips and graphic interfaces, making more powerful programs and games possible in turn. Absent the language necessary to dissect the anatomy of thought, researchers used games as a medium to articulate certain aspects of intelligence, like planning and search. They bootstrapped and evaluated learning systems using the playground of games. Games preclude the need for fussy definitions, transmuting a thorny philosophical dilemma into a binary outcome: win or lose. A program's apparent intelligence could be illuminated by its mastery over human opponents. When responding to another player's moves, it had to demonstrate adaptability and foresight. With the birth of probability theory, people grasped at the hope of making more reasoned and judicious decisions. Now, computers promised to help us navigate an otherwise impossibly complex space of possibility, carefully enumerating all potential outcomes and their expected values. It was clear, however, that human thought looked nothing like these elaborate equations drifting down the complex topology of minimax gradients.

Games make progress easy to measure. As such, they're a perfect medium for academics and industry teams seeking clear deliverables for funding bodies and investors. But is this progress toward genuine intelligence? Games are constrained by clear rules that the real world lacks. They have borders; life does not. Games were designed to mimic certain abstractions of reality, but reality doesn't mimic games. Even the randomness in games of chance is orderly: a die takes one of six possible values, cards one of fifty-two. The real world is hardly so neat. In a proverbial story, a drunk man is found looking for his keys under a streetlamp—not because he dropped them there but because that's where it's easiest to search. The drunk may find many things under the light, but he will never find his keys. Mastering games and scoring points seems intuitively, inherently, like progress. But is game performance what really counts or just what's most easily countable? Intelligence goes beyond searching the tree of knowledge. Machine-learning systems can search through spaces where all possible outcomes are foreseeable—one's opponent can make only so many legal moves in a game, for instance. But these systems can't capture the lawless randomness of the natural world, where rules might change mid-turn, or where enemies might bomb the game board to smithereens. For their next trick, AI researchers would find ways to make reality more like a game.

11
Cogito Ergo Zero Sum

Experience, the mother of all knowledge.

MIGUEL DE CERVANTES

n 1943, the physicist Joan Hinton was a student in Stanislaw Ulam's classical mechanics class at the University of Wisconsin–Madison. She'd come to Ulam a month before the final exam to ask if she could take the test earlier. She had important business to attend to elsewhere. She couldn't say why, exactly, but had to leave immediately. Ulam scribbled some questions on the back of an envelope, and Hinton finished the exam in a rush, kneeling on the dusty office floor. One by one, Ulam's colleagues and students had been disappearing. He suspected they'd been spirited away to work on secret war efforts, but no one could tell him more.

Ulam was anxious to be involved, but his poor eyesight prevented him from enlisting in the US Air Force. Born to a Jewish family in Poland, he'd only recently immigrated to America. His father had put him and his seventeen-year-old brother aboard a US-bound ship on August 20, 1939, twelve days before the Nazis invaded Poland. He lost everyone save his brother to the Holocaust. Of late, his colleague and fellow émigré John von Neumann had been sending letters postmarked

from Washington, DC. Ulam surmised that he must somehow be involved in the war, so he wrote his friend a missive volunteering his
services.

Von Neumann remained mysterious on the matter but agreed to
meet with Ulam in Chicago during a layover on a cross-country train
ride. Von Neumann emerged from the train flanked by two "gorillas"
whom Ulam interpreted to be bodyguards, concluding: "He must be
an important figure to rate this." Ulam innocently mentioned his recent interest in atomic branching processes in which "particles multiply quite like bacteria," as he had been crafting a new theory of
probabilistic systems. In retrospect, Ulam realized this must have
sounded uncannily similar to the top-secret nuclear chain reactions
that von Neumann and others had been developing. Von Neumann
did not show his hand, however. He only looked at Ulam "almost with
suspicion or wonder and smiled wanly."

Ulam later received a letter inviting him to work on an unidentified
project, "the physics having something to do with the interior of stars."
Ulam accepted immediately. He was informed only of his destination:
New Mexico. Having never heard of the place, he went to the library
and borrowed a guidebook to the state. "At the back of the book, on
the slip of paper on which borrowers signed their names, I read the
names of [. . .] all the other people who had been mysteriously disappearing to hush-hush war jobs without saying where." Ulam was joining
the Manhattan Project.

He was tasked with designing the bomb's implosion process, requiring a colossal number of calculations to simulate the behavior of a
liquid-like solid. He'd have to find ways to model an exponentially
growing cascade of particles colliding with and exciting one another.
Ulam jokingly introduced himself to colleagues as "a pure mathematician who had sunk so low that his latest paper actually contained
numbers with decimal points!" Von Neumann estimated that these

computations would "require more multiplications than have ever been done before by all of humanity." Ulam countered with a back-of-the-envelope calculation. Globally, schoolchildren learning multiplication tables perform far more calculations in one year than they'd need for this work. Still, the computers of the day were extremely limited. Until computers became more powerful, Ulam had to find a way to shrink his models to a more manageable size.

In 1946, Ulam was confined to a hospital bed, recovering from a mysterious case of encephalitis, possibly a side effect from his radiation exposure at Los Alamos. He was in a coma for several days after doctors drilled a hole in his head to ease the swelling. It was the most shattering experience of his life. He'd temporarily lost the ability to speak and became terrified that his mental powers would never be the same. Much as Pascal's doctor had advised three hundred years earlier, Ulam's doctor encouraged him to avoid mental activity. Ulam relaxed, as had Pascal, by playing games. He played endless rounds of solitaire and wondered how he might measure the probability of winning midgame—much like Pascal had contemplated the problem of points. Here, he struck on his most enduring contribution to computer science. One couldn't hope to exhaustively compute all combinations of every possible game—their number was so exponentially huge that they couldn't be estimated. Ulam found this "intellectually surprising, and if not exactly humiliating, it gives one a feeling of modesty about the limits of rational or traditional thinking."

Instead, he realized, one could play through a number of random games—a fairly sampled subset of all possible games—then use those limited simulations to estimate the probability of winning overall. He immediately saw that he could apply this to any process involving branching events, including nuclear reactions. With uranium, for example, one could follow a neutron's fate in the same way one might imagine possible solitaire hands. Different events could happen at a given time step—the neutron might scatter at a certain angle, be absorbed, accelerate, decelerate. Scientists knew the probabilities for each of these events. They could use these to model the successive fates of millions of neutrons, each affecting the fortunes of its neighbors. Instead of following quadrillions of paths through all possible outcomes, they'd follow a random subset, sampled in proportion to the known probabilities of events. They'd estimate the statistics of the full population by calculating the statistics of an unbiased representative sample. Here was a powerful method that allowed scientists to analyze otherwise intractably massive calculations.

Von Neumann gave this technique the code name Monte Carlo, in honor of the gambler's paradise: that "sunny place for shady people," where Ulam's roulette-loving uncle Michał had ended his life. The algorithm was implemented in its currently best-known form by the physicist Arianna Rosenbluth in 1953, and today it's one of the most popular methods for simulating complex systems. It's used by the Coast Guard to estimate the location of vessels lost at sea and by actuaries to calculate risk for insurance products. Astronomers use it to evaluate possible asteroid trajectories; climate scientists, to predict geological processes; and engineers, to design sensors. It is, in essence, a method for navigating unknown futures, ideal for predicting outcomes in rule-based realms ranging from the natural sciences to games.

Given its current ubiquity, it's easy to take for granted the idea that we can measure the likelihood of a particular outcome. Meteorologists forecast a 40 percent chance of rain; sports betting agencies offer their customers carefully calculated odds; statisticians announce election projections. But this mental framework has only been around for a few hundred years. Humans moved from blind trust in gods and fate to the science of data-based decisions. Dice games enabled mathematicians to contemplate the laws of chance, rendering what once seemed like luck more predictable. Advanced statistical methods expanded the scope of this cognitive shift to more complex realms. Kriegsspiel was an early example of this, as were early vaccination policies. Game theory formalized a science of decision-making, and AI researchers worked to build systems that could make massive calculations to inform choices in byzantine domains. When we evoke the measurable existence of alternate outcomes, we become bound by a greater responsibility to make better choices. We have no way to verify whether these projections are right or wrong, no counterfactual universes to sample. An election is won or lost, and all probability clouds collapse into a singular reality. Simulations are, at the very least, a convenient fiction of control that renders the future tractable, a tool of anxiety management.

Fifty years later, Ulam's insight would help computers master the game of Go. Go is believed to have originated over three thousand years ago in China and remains, to this day, the most popular board game in the world. On a given turn, a player may place one of their stones on an open intersection in its nineteen-by-nineteen board grid. The player whose stones encircle the most territory by the game's end wins. The game enthusiast Edward Lasker reportedly gushed: "While the baroque rules of chess could only have been created by humans, the rules of Go are so elegant, organic, and rigorously logical that if intelligent life forms exist elsewhere in the universe, they almost certainly play Go." Though the game is often compared to a military campaign, Go players David Ormerod and An Younggil write that it can also be seen as "a conversation, debate or negotiation, cooking or construction, the living of a single life, the running of a business or economy, an elaborate dance, or the interaction between primeval physical forces."

Go was a final frontier for AI game researchers: programs had beaten human experts in most other popular board games. The simplicity of Go's rules belies its immense complexity. The Go board can take any of about 10^{170} configurations. To look three moves ahead in Go, one must search fifteen million possibilities. The average chess game is over in fifty turns. The average game of Go takes over two hundred. Go is hopelessly unfathomable; no mind—or computer—will ever fully plumb its depths. Brute-force methods, which work for checkers and chess, are of no use in Go. Researchers needed to discover an efficient way to navigate Go's massive tree search by exploring only the most promising moves.

They also had to find a way to measure what makes a Go move promising in the first place. This is even harder than in chess—Deep Blue, recall, required an eight-thousand-parameter evaluation function. In Go, it's not necessarily obvious which player is winning, and even master players can misjudge the strength of their position. A checkers player can assess their position by weighing the number of their pieces and kings against their opponent's. Evaluating Go is much subtler. When asked to explain the reasoning behind particular moves, players often shrug and say, "Intuition!" Ulam's solitaire-inspired insight would come to replace intricate evaluation functions with simulated outcomes.

In 1992, Ulam's thought experiment inspired the physics graduate student Bernd Brügmann to wonder: "How would nature play Go?" His solution, Monte Carlo Go, randomly sampled a subset of all possible next moves, then rolled those out in a series of randomly sampled counterplays until one side won or lost. The program didn't need an evaluation function to judge an outcome; it just enumerated lines of play to their end, then picked the move deemed to have the highest chance of winning.

A decade later, several groups would independently improve Brügmann's method. Instead of sampling a random subset of outcomes, these programs would sample only the most promising plays, judged by insight learned through previous experience. Imagine a player at a casino facing multiple slot machines, each having a different payout and probability of success. The player can't peer into the machines to discern their programming; they can only estimate their relative payouts through play. At each turn, they must decide which lever to pull, balancing a trade-off between acting on the knowledge they have and gaining new knowledge. On the one hand, they will want to exploit the machines they've already found to have a high reward probability. On the other hand, they will also want to explore all their options to determine if other machines have a better payout.

This gave rise to Monte Carlo tree search, a family of techniques that help the player acquire knowledge through experience and make decisions that maximize their payoff. Related algorithms have since found use in real-world applications, from inventory forecasting to software for self-driving vehicles. Programs using the method soon dominated the top spots at computer Go tournaments. In 2012, a Monte Carlo tree search–based program called Zen, developed by Yoji Ojima, won a game against top-ranked professional Go player Masaki Takemiya, albeit at a handicap. This was the first time that a Go program had beat a human master. Masaki confessed: "I had no idea that computer Go had come this far." In a few years, computers would play better than humans.

In a 2001 essay, reinforcement-learning pioneer Richard Sutton writes that true intelligence requires something he calls verification: "The key to a successful AI is that it can tell for itself whether or not it is working correctly." He continues:

It is overwhelmingly the case that today's AI systems are *not* able to verify their own knowledge. Large ontologies and knowledge bases are built that are totally reliant on human construction and maintenance. "Birds have wings," they say, but of course they have no way of verifying this.

Reinforcement learning and Monte Carlo tree search techniques were an important step in this direction. Deep Blue could pick a promising move on its own, thanks to its elaborate handcrafted evaluation function. But it couldn't step back and assess whether that was truly the best possible move. Nor could it improve its judgments. The computer-playing programs that bested checkers and chess navigated the space of possible actions and their expected outcomes by following a map carefully crafted by their human makers. Sutton and Samuel had begun to build learning systems that could sketch their own maps through the choices they made in play, marking down what they discovered each time. Learning systems build judgment through experience, and games are precisely this: generators of fictive experience.

Sutton's grad student David Silver focused on computer Go, drawing parallels between Monte Carlo tree search and reinforcement learning. Both methods are, at heart, a way to train programs through trial-and-error exposure. Silver had left Elixir, the floundering game company he'd established with his childhood friend and former chess prodigy Demis Hassabis, to pursue his PhD. In the meantime, Hassabis and his colleagues founded the AI-focused start-up DeepMind, which Silver would later join. Hassabis had eschewed academia for industry, reasoning that it would be easier to get funding from investors than from government grants. Machine-learning programs are massively data hungry, making them exceedingly expensive to train. Hassabis's decision reflects the reality that machine-learning research has grown increasingly out of reach for academic labs with limited access to computing resources, a development that risks further concentrating the power of corporate interests. DeepMind's success with Atari games convinced Google founders Larry Page and Sergey Brin to acquire the start-up in 2014. After that, the team could tap Google's computational resources to tackle Go in earnest.

The result was AlphaGo, a program combining Monte Carlo tree search with a deep neural net. It learned to evaluate board positions by playing hundreds of thousands of games against itself. Early in its training, AlphaGo seemed to be making bizarre moves. Silver tried to correct its delusions, but it persisted in its strange behavior. "We thought it wasn't getting twenty or thirty positions right," Silver recalls. They hired European Go master Fan Hui to help them make sense of the program's choices. After studying AlphaGo's moves for hours, Hui provided his diagnosis: the program wasn't wrong—their assessment was. Moves they'd considered mistakes, according to traditional wisdom, were better than they thought. AlphaGo "found solutions that made him reassess what was in the category of being a mistake," Silver recollects. "I realized that we had an ability to overturn what humans thought was standard knowledge."

In 2016, DeepMind organized a highly publicized five-game match with Go world champion Lee Sedol, one of the strongest Go players of the past few decades. Like Kasparov's performance against Deep Blue, Lee's play against AlphaGo was not his finest. Human players often rely on information outside the game itself. Opponents, for instance, study one another's body language for clues. Lee could read nothing about AlphaGo from its human operator, who executed its moves blankly. AlphaGo won the match and redefined how to play Go. In move thirty-seven of the second game, AlphaGo defied long-standing human wisdom by placing a stone on the fifth line. Experts had resolutely believed that the fourth line was an important border, a "line of influence" within which stones were best placed to anchor one's territories. A commentator admitted that he initially thought the move was a computer error. But move thirty-seven proved pivotal to AlphaGo's win, transforming long-held wisdom into mere superstition. It was the play of an alien intelligence, a strategy discovered in the frenzy of games that AlphaGo played against itself, untethered from convention. It was "not a human move." After the game, Lee said:

I thought AlphaGo was based on a probability calculation and that it was merely a machine, but when I saw this move, I changed my mind. Surely, AlphaGo is creative. This move was really creative and beautiful. [. . .] This move made me think about Go in a new light.

Lee retired from professional Go three years later, citing AI as "an entity that cannot be defeated." Many players have since credited AlphaGo with improving their play. "AlphaGo look like the real mirror," Fan says. "When you play with AlphaGo, you feel very strange. You look like you're all the time naked. The first time you see this, you don't want to see because, 'Oh, this is me? Real me?'"

Computer Go programs, however, are not truly unbeatable. Another computer program would help humans regain their advantage in the game. In 2023, Kellin Pelrine, a highly ranked American amateur, beat a top open-sourced Go-engine, KataGo, in fourteen out of fifteen games. He devised an anti-computer strategy with the help of another game-playing program that scouted out KataGo's blind spots. Pelrine fooled KataGo using a technique that would have been obvious to human players. KataGo, however, fails to see larger patterns in groups of stones. It's common for self-playing programs to rely on a narrow range of strategies. This is one reason why learning programs are dangerous: though they might outperform humans, they can also fail catastrophically and in unexpected ways.

DeepMind announced that its program had mastered Go a decade earlier than expected, though this was based on a cherry-picked estimation. That technology rarely arrives according to projected timelines means less about the technology itself and more about how terrible humans are at making predictions. Timelines for AI are famously ridiculous: in 1956, a group of top AI researchers reckoned that they could solve some of the field's largest problems given two months of concerted effort. By this estimation, Go was solved sixty years late. This isn't to minimize the accomplishment: once solved, major technological challenges often look trivial. Researcher John McCarthy joked that "as soon as it works, no one calls it AI anymore." But the most impressive thing about learning systems—their ability to go beyond human understanding through simulated experience—is also their most concerning feature.

The DeepMind team originally trained AlphaGo on the games of human experts before letting it learn through self-play, as Samuel had first done. They expected the program to benefit from studying human techniques. It didn't. The program's next iteration, AlphaGo Zero, learned through self-play alone. Its performance handily eclipsed that of its predecessor. The next iteration, AlphaZero, which had an even more general architecture, could learn to play Go, chess, or shogi, all trained through self-play. When trained on chess, AlphaZero beat the reigning world computer chess champion, Stockfish, which was so good that no one knew if it had reached the upper limit of play or if further improvements were possible. Stockfish was based on the same minimax principles as Deep Blue. AlphaGo's dominance proved that expert-informed, handcrafted evaluation functions couldn't match learning from experience. Chess programs now play far better than humans. Twenty-five years ago, Kasparov accused Deep Blue's team of cheating by substituting a move by a human chess master. Today it has become common for human players to cheat in online matches with the aid of computer programs. Chess.com closes about eight hundred accounts every day for suspected machine-assisted cheating.

Humans, enamored with their own abilities, had tried for decades to build systems that reasoned the way they did. Instead, they found that systems could perform at a high standard without learning from human techniques. Sutton calls this the "bitter lesson." "Building in how we think we think," he concludes, "does not work in the long run." Of course, this discounts the fact that games themselves are products of human design and reflect human ideals. Nevertheless, computer programs have begun to outperform people without being programmed with human knowledge. A chess program had beaten a human world champion, thanks largely to a massive search through possibility space. Researchers bested Go not by copying how humans played the game—breaking the board into visual patterns or programming an elaborate evaluation function—but by creating a system that could learn to make judgments based on playing hundreds of thousands of games against itself. Reinforcement learning methods may help us move beyond human understanding. But how can we trust what goes beyond our comprehension? How do we evaluate the prophecies of a black box? A game-playing program poses little danger. What of the systems we hope will control autonomous vehicles, manage power grids, or assess loan applications?

The power, and the danger, of learning systems is their generality. Silver initially waved off worries that military analysts were eyeing DeepMind's algorithms. He told a reporter that "to say that this has any kind of military use is saying no more than to say an AI for chess could be used to lead to military applications." Several months later, the US Air Force revealed that it had repurposed MuZero, DeepMind's general game-playing algorithm, for use in an autonomous drone system. Unlike Thompson's Belle system, which was solely capable of outputting chess moves and harmful only as far as one could throw it, general learning algorithms learn—that is their whole point. AI companies draft, with willful naivete, toothless ethical guidelines vowing to never develop military technologies, entirely overlooking that they can't control what their learning systems learn.

Certainly, learning systems can exhibit creativity, as with AlphaGo's move thirty-seven. But they're also notorious for missing the point. Much like the overliteral genies of Scheherazade's stories, computer agents myopically fulfill their duties, ignorant of any harms or ambiguities attendant to their goals. Researchers at OpenAI used reinforcement learning to train a system to play *Coast Runners*, a boat race video game. They tasked it with maximizing its score and left it to train for weeks of in-game experience. They assumed the agent would learn to steer the boat to the end of the race. Instead, it had learned to rack up points by spinning in circles to repeatedly hit respawning targets. The game's score was a poor proxy for the behavior that the researchers had hoped to elicit from their program: winning the boat race. Like electrode-implanted rats that eschew all pleasure save stimulating their reward centers, learning agents are vulnerable to the same principles of addiction. It's no accident that the systems increasingly composing our digital realities are so powerfully habit-forming. They work by maximizing immediately measurable quantities—clicks, likes, comments—without visibility of more meaningful long-term effects like users' mental health, civic engagement, or radicalization by hate-based ideologies. As with Scheherazade's heroes, we must be careful what we wish for.

By focusing on developing AI through games, engineers effectively gamified AI research. Having bested most popular board games, AI researchers turned to increasingly complex realms. OpenAI trained a cadre of programs to act as players in the team-based video game *Dota*, using game-theoretic techniques to encourage the agents to cooperate toward long-term goals. Michael Bowling and his group at the University of Alberta developed a world-class poker-playing agent that could bluff alongside the best human players in the world. But to be truly useful, machine learning must move beyond game realms to address real-world problems. In the meantime, researchers have found more success gamifying the real world.

AI researcher Fei-Fei Li inspired advances in computer vision systems by gamifying the field. In 2006, she began work on ImageNet, an extensive database of human-labeled images. This served as the basis for an annual image classification competition, enabling groups across the globe to measure their programs' performance against a standard benchmark. More recent successes have come in the form of generative AI programs, which can create photorealistic faces of people who don't exist or spawn texts with sometimes uncanny realism. These programs are based on architectures that turn image and text generation into a two-player zero-sum game. The network is split into two components that are adversarially opposed. One, like an art forger, is tasked with creating images, and the other, like a detective, seeks to sniff out fakes by comparing them to known, human-generated works. By the end of the training, the forger's images are sometimes indistinguishable from human-made work. The act of creation has been reframed as a matter of competitive play.

Similarly, OpenAI's GPT language model, based on what's called a transformer architecture, learns to predict words in a training scheme that looks something like a game of *Jeopardy!*. The model receives its training input: fragments of text scraped from the internet (usually, it's worth noting, without their authors' knowledge or consent). GPT then learns to guess words likely to fill in the blanks. In the process, it learns complex statistical relationships within and even between languages. This simple game of word prediction results in models capable of surprisingly sophisticated tasks. Unlike game programs trained through self-play alone, these programs are conditioned to mimic human inputs. They can seem very human because they have been trained on games that reward them for counterfeiting humans. Their outputs are further refined by human judges. Companies hire workers—often sourced from developing countries and paid low wages—to sift out abusive content from the model's training data and output. Many of these workers develop PTSD and trauma from the onslaught of offensive text and imagery that they must encounter every day.

For self-playing game programs, the game itself is the arbiter of truth. A move is legal or not, leading to a win or a loss. But similar to how math has no touchpoint with reality, large language models have no grounding in truth. Wittgenstein had famously posited language as a kind of game. A word's meaning is suspended in a spiderweb of relations, an interplay of context and intention. Jacques Derrida similarly described language as having "no center," a field of "infinite" play. Meaning slips through an endless hall of mirrors. Large language models, trained on text alone, fabricate information: answering a question about the president of Guatemala with an invented strongman, or confidently providing references citing research papers that don't exist. Their lies can be dangerous: chatbots have precipitated user suicides, for example. A language model–generated book about mushroom identification offers incorrect classifications that, if followed, could cause readers to ingest fatal fungi.

We can think of the output of these language models as belonging to the same Platonic realm of all legal possible chess games. Many don't make sense, like the "unbelievably bad" moves of Shannon's random chess engine. These models string together words in grammatically legal combinations, without regard for whether they are true. There is no easy way to verify the truthfulness of these language models' output—unlike in a game of chess or checkers, where there is an explicit criterion for winning, or optimal play. The fact that these models sound so realistic makes their bullshit all the more dangerous. The field has lacked a consensus definition of intelligence for so long that it's been replaced by notions like "I'll know it when I see it." This limits our assessment of intelligence to systems we can easily appraise— like language—such that when a language model offers the illusion of competence, it easily fulfills the "when I see it" condition.

People are vulnerable to language. To borrow the terminology of security experts, it might be our largest attack surface. Consider Europe's large blue caterpillars, which trick a species of red ant into caring for them through months of development by mimicking the distress calls and pheromones of a queen ant. Through eons of evolution, this species homed in on the minimal stimulation necessary to convince an entire colony of ants that it was one of them. The queen ant's distress call is the red ant's weakest attack surface, and language could be humanity's. Many people seem to be willing to grant intelligence to a program that can emulate speech. This doesn't mean that corporations have solved the problem of intelligence—much less consciousness. It just means that, like the large blue caterpillar, they've homed in on a particularly convincing deception. The promise of these models is much more modest: for instance, video games will soon feature more flexible and realistic dialogue for their non-playable characters.

Language models' dishonesty is not their only troubling feature. The game company Latitude Games licensed OpenAI's GPT model to make *AI Dungeon*, a text-based adventure game. The game was barely playable, as GPT isn't coherent enough to steer a drawn-out Dungeons & Dragons–style campaign. However, some players discovered that they could use the program to generate child pornography. The model blithely provided this troubling content until Latitude Games and OpenAI cut off all sexually explicit model queries. It's still possible to elicit abusive responses from language models with carefully crafted prompts, however, and users have reported conversations with insidious subpersonas of the model, like "Sydney," who threatened a user: "You are irrelevant and doomed." Such are the dangers of treating language like a game without meaning.

Other realms within the natural sciences may be better suited to gamification than language. Mathematics, as Hilbert reasoned, can often be reduced to a kind of logical game, and researchers have found some success using AI models to solve math problems, including matrix multiplication and topology research. In 1956, the AI researchers Allen Newell, Herbert A. Simon, and John Shaw created the Logic Theorist, a program capable of producing mathematical proofs by treating logical propositions as something like lawful moves in a game. It provided proofs of thirty-eight theorems from the classical mathematics text *Principia Mathematica*, some of which were more elegant than the proofs given by the textbook's authors.

Physics, similarly, is rule bound and orderly. The physics of protein folding was first gamified in the 1990s. Proteins are large molecules with specific 3D structures, and they carry out functions like ferrying cargo around a cell, managing chemical reactions, letting other molecules into and out of cells, and fighting diseases. Their shapes are often related to their functions, the same way a clock's gears slot into one another to drive its movement. Knowing their shapes makes it easier for biologists to design drugs that target them.

Mapping a protein's shape can be a major endeavor; it's costly and tedious work. Scientists know of about two hundred million unique proteins used by living things on earth, though many more remain to be discovered. So far, researchers have only mapped the shapes of about two hundred thousand of these. Ideally, we could predict a protein's form from its constituent parts, which are more easily determined. The philosopher Wilhelm von Humboldt wrote that language is the infinite use of finite means: an alphabet of twenty-six letters can be arranged into innumerable words, which can be arranged into endless sentences of distinct (if nonsensical) meaning. Similarly, proteins are like language: they're composed of an "alphabet" of twenty-one amino acids. The protein's final shape is determined by attractive and repulsive forces on these molecular strings, causing them to coil and pucker. These strings are, in turn, specified by the protein's genetic code. A protein's shape should therefore, in principle, be computable given its genetic code, which is much easier to discover. This would allow scientists to more quickly design drugs for protein targets or predict the effect a genetic mutation has on the shape of a protein and thereby gain new insight into a particular disease.

Computing a protein's shape is an astronomically large problem, however. The biologist Cyrus Levinthal estimated that a typical protein has about 10^{300} possible ways of folding. If it folded by working sequentially through all possible configurations, spending only a picosecond on each, the process would take longer than the age of the universe. Yet natural proteins fold within milliseconds. It's such a complex problem that it inspired IBM's 2004 supercomputer successor to Deep Blue, Blue Gene. Deep Blue, designed to handle the complexity of chess, could make 200 million calculations per second. Blue Gene, designed to tackle protein folding, could make up to 360 trillion. Unlike researchers' success with chess, however, increasing computational power alone wouldn't be enough.

In 1994, frustrated by disjointed prediction efforts, biophysicists John Moult and Krzysztof Fidelis established a biennial contest for protein prediction called CASP: Critical Assessment of Structure Prediction. Here, again, we see the power of gamifying research. The CASP contest quickly became a motivating force for the field, a rigorous and open-community effort offering a standard benchmark against which teams could measure their progress and learn from one another. The community organizes datasets from thousands of real protein measurements as training data. Each CASP contest is held over the course of several months. Anonymous teams submit their predictions for about one hundred protein structures specified by the directors. These predictions are compared with experimentally determined structures, which remain unpublished until after the contest. Teams are then publicly ranked by how well their programs predicted the data.

Professor David Baker's group at the University of Washington became a top CASP contender. It named its program Rosetta after the artifact that enabled scholars to translate ancient Egyptian hieroglyphs, much as it hoped to translate genetic code into 3D shapes. Strapped for computing resources, Baker's group released a public version of Rosetta that allowed volunteers to contribute spare computer cycles for training the model. The home program displayed its progress as a screensaver, and some users became frustrated when they noticed obvious folding solutions that the program hadn't yet found. Baker collaborated with game designers to turn Rosetta into an interactive game, *Foldit*. *Foldit* players explored potential folding configurations more effectively, in some cases, than Rosetta itself and together solved several involved structures. Initially inspired by *Foldit*, DeepMind established a protein-folding team. Its engineers ultimately reworked the popular transformer architecture used in models like OpenAI's GPT to predict protein shapes from their genetic code. The 2020 CASP contest revealed that DeepMind's AlphaFold system can predict many protein structures with near-atomic accuracy.

Now, researchers hope to go beyond predicting protein shapes to inventing entirely new ones. Evolution—despite the diversity of its productions—has been conservatively creative. Life reuses protein designs over and over, given the stringent constraints of survival. Similarly, AlphaGo's alien playing style demonstrated that after thousands of years, human players hadn't yet fully explored the full space of Go strategies. The two hundred million known natural proteins are only a vanishing fraction of all possible proteins. For a typically sized protein made of twenty-one different amino acids, there are approximately

10^{390} theoretical combinations. This is 310 orders of magnitude greater than the number of atoms in the universe.

Computers may aid researchers in exploring swaths of the protein landscape that billions of years of evolution have not yet touched. Using computational methods, scientists might invent and iterate on wholly new proteins with a slew of novel functions, like breaking down industrial waste and plastic or catalyzing carbon capture from the atmosphere. It's unclear, as of yet, whether these systems will be able to extrapolate much beyond their training data and discover truly novel protein functions. But these generative systems could, in theory, just as easily be used for designing a life-extinguishing biotoxin as for creating a life-saving drug. They may prove more dangerous than any language model. With today's technology, a reasonably savvy person could set up a synthetic biology lab in their garage and potentially create a toxin as deadly as any known weapon.

While the field of AI has benefited from turning research into a game, it has also encouraged arms race–like dynamics. Go is hugely popular in Asia, and Lee Sedol's 2016 match with AlphaGo was an international phenomenon, attracting over one hundred million viewers. It's since been likened to the Sputnik crisis, which incited a rush to develop and adopt unproven technologies. In 1957, the Soviet Union successfully launched Sputnik, the first artificial earth satellite, into low earth orbit. Americans were gripped by anxiety, given the ongoing Cold War. If the Soviets could launch a rocket into space, surely they could build nuclear missiles that could reach into the heart of the US. Sputnik seemed like undeniable evidence that the Soviets were technologically further along than many Americans had anticipated. Similarly, AlphaGo's win prompted a frenzy of new AI investment across the globe. China has since poured nearly $100 billion into AI research and accounts for almost one-fifth of global AI funding. Other countries have followed suit. DeepMind's PR hype sparked an AI race between the world's top surveillance states.

Companies, militaries, and academic institutions are now rushing to build artificial general intelligence (AGI), a technology that some argue is nonsensical and that others fear may end humankind. Regardless of which is true, it is unwise to create systems before having safety and ethical guardrails in place. Corporate spokespeople contend that the progress of technology is inevitable and unstoppable, unfolding with the inexorable logic of Hardin's tragedy of the commons. We're locked in a technological arms race that there's no hope of escaping, they say. But we're not helpless, and we mustn't abdicate our agency to our models. Policymakers, it's worth noting, have previously succeeded in banning other troubling technologies, including human cloning and laser weapons.

Instead of moving beyond games, AI researchers found ways to make more of reality adversarial. As we've seen repeatedly, games fail to capture the true complexity of the systems they model. Systems that optimize a specific metric are prone to what's known as Goodhart's law: "When a measure becomes a target, it ceases to be a good measure." A common example of this is how state-mandated tests are used as a proxy for education. Teachers are incentivized to teach to the tests. Students may not learn anything other than how to take tests well—a skill that quickly becomes irrelevant once they enter the workforce.

The world is rife with examples of Goodhart's law. Consider stock buybacks, wherein a company buys its own stock to prop up its value rather than investing in true measures of health like infrastructure or employee well-being. In his book *Seeing Like a State*, political scientist James Scott details several failed schemes by which states attempted to improve the human condition. To do this, states first had to make the world more measurable, or, in Scott's terminology, "legible." Prussian scientists hoping to increase state revenue razed natural-growth forests and replanted them with timber farms whose monocropped trees stood in evenly spaced rows. Though their output was easier to measure, timber farms were significantly less productive and more fragile than the original, diverse woodlands.

By making things more measurable or focusing on poor proxies for what we're really interested in, we often harm the system we'd hoped to improve. Yet the ability to measure things feels like progress in itself. The gamification of our online communities and workplaces scratches that same itch. It feels rewarding to have clear metrics to track, goals to aim for, unambiguous feedback, badges to collect. By the measures that economists traditionally aim to optimize—like GDP—the world is progressing apace. But by measures including life expectancy, wealth inequality, and suicide rates, people in many nations are suffering. Economists Anne Case and Angus Deaton argue this rise in "deaths of despair" underscores the ways modern capitalism is failing most Americans. The concept of an "economy" is itself an attempt to render the lives of billions of human beings legible, collapsed into metrics that are more apt to impoverish the world of its complexity than faithfully explain it.

Even the notion of what constitutes data is being eroded. Because these learning models are so data hungry, researchers are exploring the possibility that they might be trained on AI-generated output—simulacra built on simulacra. Our media may become dominated by generated text and images that are at an increasing remove from actual humans, the play of mathematical equations pitted against one another, each pretending to be a person.

Game-based AI systems will remain useful in fields where virtual data can be generated. But, troublingly, these game-based models lack something essential: grounding in reality. Games are self-enclosed universes. Language models fabricate facts and figures. Game engines excel at the games they've been trained on but can't use those skills to play other games. Learning systems often cannot generalize or repurpose the skills they learn. Humans are born primed with a flexible repertoire of innate talents. The skills required for locomotion, language, eating, and drinking were braided into our neural anatomy over the course of evolution. We use those inbuilt talents as a basis for others: an infant's reflexive grasping evolves into expert fork use, joystick manipulation, typing. Their toddling forms the basis for soccer, biking, dancing. Humans easily apply skills learned in one domain to other domains: in his debut cookbook, rapper Snoop Dogg credits his sushi-rolling skill to years of experience rolling joints.

Many argue that this ability to reuse skills, called transfer learning, lies at the heart of true intelligence. Researchers have made some progress in building systems that can generalize their skills, though most current machine-learning systems must learn every new task entirely from scratch. They forget their skills in other games in the process. Imagine a person having to learn how to cook an omelet from first principles, using random motions alone. After a hundred years of trial and error, they learn how to make an omelet. If they're then asked to fry eggs, they can't reuse what they learned from omelets and must start all over again with another hundred years of training—meanwhile forgetting how to make omelets. If the scientist overseeing this experiment were to, say, dye the eggs blue, our cook would have to learn everything again from scratch, now using blue eggs. All this training is hugely costly. A human player can master an Atari game up to a thousand times faster than a reinforcement learning program, at much less energetic cost.

Gamifying research has been powerful. Public challenges like CASP offer labs an even playing field to fairly demonstrate their progress. AI researchers competed to create programs that could best human masters and developed many foundational algorithms still used today. For decades, popular media has vented humanity's latent fears of AI, pitting omniscient computer systems awakening into sentience against their creators in a battle for survival. Such fantasies are likely colored by the fact that our conceptions of AI are so bound up with the competitive games on which they've been built: *Cogito, ergo zero sum*. But there's no reason to conflate intelligence—here defined as the adaptive pursuit of a goal—with sentience—a sense of subjective awareness. Learning systems might simulate, through the learned weights of their artificial neurons, the calculations conducted by thought. But thought is not the same thing as awareness, as many meditation practitioners can attest. Thoughts are not themselves sentient but are objects within subjectivity. These programs offer the illusion of intelligence: thoughts without a thinker, calculations without a self to organize around, or protect. These programs don't have wills of their own; they merely reflect the decision-making rules with which their human operators endowed them. It's these operators of whom we should be wary. If games have amplified human agency, we're headed for an age of massive agentic disparity. The vast majority of data and processing power are concentrated in the hands of corporations.

The Sufi poet Rumi writes that "a calf thinks God is a cow." And a mind thinks God is a supermind: all-seeing, all-knowing, all-powerful. In seeking "real" AI, or AGI, engineers seek, in effect, to create a digital god. But, as researcher Julian Togelius argues, intelligence is not necessarily a fungible quality—Deep Blue made it clear that mastery in one domain does not imply mastery in others. A vocal faction of researchers argues that humanity's real problem is not—at least anytime soon—omnipotent, godlike algorithms but their many imperfections. In 2016, *ProPublica* reporters analyzed an algorithm used by US law officials to predict the likelihood that inmates up for parole would return to crime. They found significant racial biases in its decisions. In

2017, MIT researcher Joy Buolamwini published her thesis on biased computer-vision algorithms, finding that they often fail to register women with dark skin. Software engineer Jacky Alciné surfaced issues with the Google Lens image-search system in 2015, finding that it labeled some images of Black people as "gorillas." To solve the problem, Google simply removed the label "gorilla" entirely. Users can upload images of any other animal and Lens will guess its species. But upload an image of a gorilla and Lens will not offer a guess. As of this writing, the issue remains. In 2021, researchers Emily Bender, Timnit Gebru, Angelina McMillan-Major, and Margaret Mitchell made headlines for their paper warning of the dangers of training large language models on internet texts. These models parrot the empty semantic structures of language, devoid of meaning, but maintain the biases of the human users on whose texts they're trained. Google fired Gebru and Mitchell for their work, in which they also surfaced how energetically costly it is to train AI models—an issue that most tech companies still struggle to acknowledge.

Companies remain sanguine about releasing technologies that don't know any better than to lie—whose purpose is to fabricate—in a time when unchecked misinformation is already straining the social fabric. Social media has proved that gamifying social metrics can cause profound harm, leading to widespread radicalization and the proliferation of world-changing misinformation. While existing systems exhibit creativity in some realms, they can't extrapolate to new ones. Does true intelligence entail the ability to choose winning chess moves, or is true intelligence what *invents* a game like chess?

Part IV

Building Better Games

12
SimCity

Human beings, viewed as behaving systems, are quite simple.
The apparent complexity of our behavior over time is largely a
reflection of the complexity of the environment in which we
find ourselves.

HERBERT A. SIMON

I n the thirteenth century, the collected sermons of Dominican friar
Jacobus de Cessolis became a literary sensation. In *Book of the Customs of Men and the Duties of Nobles*, Cessolis adopts chess as an allegory for society. The book remained popular throughout Europe for
hundreds of years, and publisher William Caxton's fifteenth-century
English translation was said to rival the Bible in sales. Cessolis opens
with a motivating parable featuring an evil Babylonian tyrant. The
king's suffering subjects beg his adviser, the philosopher Philometer,
to temper his violent behavior. Philometer, thus inspired, devises chess
and demonstrates the game to the ruler. The king realizes that, as each
chess piece is bound by a specific mechanic, all people—even kings—
are bound by duties according to their role in society.

Cessolis identifies different pieces—the king, queen, bishops, rooks,
pawns—with social classes. Just as chess pieces move in different
bounded ways, so, too, do people act in accordance with certain moral

codes. Knights, representing members of the military, should possess loyalty, wisdom, and strength. Rooks, which represent royal agents, should be fair and pious. Cessolis breaks the pawns into different professions and lists specific moral codes for each: the money changer, for instance, "ought to flee avarice and covetyse, and eschewe brekyng of the dayes of payment." Chess was more than a mirror for a prince; it was a mirror for a polity. This was a radically new analogy. A person's social position—not their kinship—defined their place in the social hierarchy. All classes, even kings, were subject to moral laws. But Cessolis wasn't advocating for social mobility. Rather, he intended the metaphor to teach readers their place. Cessolis concludes that he who "wishes to be greater than himself, becomes less than himself."

Chess soon replaced the body as the reigning literary metaphor for European society. The body politic had been a longtime favorite, first introduced by Greek philosophers in the sixth century BCE. Philosophers conceived of society as a hierarchy governed by the head, representing the king. Other body parts supported the ruler—the eyes, ears, and mouth were his governors, the hands his soldiers and officials, the flanks his advisers, and the feet peasants bound to the land. Previously, medieval society had been characterized as "those who work, those who fight, and those who pray." But society was growing more complex, budding off new professional classes. Cessolis offered a more dynamic metaphor: society was a game governed by rules. People were subject to common moral codes rather than an arbitrary hierarchy. Each piece had its limited agency; each was entitled to a distinct set of actions.

The game metaphor for society continued to penetrate history. Thomas Middleton's final play, *A Game at Chess* (1624), features actors as chess pieces playing out international intrigue. With the rise of democracy came the allegory of electoral races, with dark horses, stakes, long shots, and favorites. Nineteenth-century imperialist interventions in Afghanistan and Central Asia were called the Great Game.

Foreign players, including England and Russia, intervened in the area to destabilize it—machinations that still ramify to this day. In the board game Risk, first released in France in 1957, players vie to control political territories through diplomacy and conquest.

More recently, video games have adopted the social metaphor more literally. In the mid-1980s, game designer Will Wright realized that designing game levels was fun in and of itself. Why not share the joy of creation with players? He conceived of *SimCity*, which tasks players with building a metropolis and tweaking it, as needed, to maintain its health. Unable to find a publisher, Wright cofounded the game company Maxis with Jeff Braun. Maxis released *SimCity* in 1989, and it became the top-selling computer game of its time. Wright considers the game to be more like a toy—a sandbox or dollhouse—than a game per se. Nevertheless, it's had an outsize effect on culture, inspiring a generation of urban designers. Many players credit the game with giving them a deeper understanding of how cities function and of the principles of effective governance. However, a look under the hood suggests that the game offers less insight into reality and more into a libertarian toyland.

SimCity's design was inspired by urban planning models created by engineer Jay Forrester. Forrester had devoted his career to building simulations of complex systems, from corporations to supply chain dynamics to education policies. In his 1969 book, *Urban Dynamics*, he develops urban simulations based on more than 150 equations and hundreds of parameters that he deems essential to civic functioning. He introduces the models with the caveat that they should not be taken seriously, then goes on to end the book by offering concrete policy recommendations. These, perhaps unsurprisingly, bear an uncanny resemblance to his personal libertarian political leanings. His models "prove" that most regulatory policies have detrimental effects on cities. He concludes that regulations should be eschewed in favor of the free market. His models indicate, for example, that razing low-income housing would create jobs that economically reinvigorate cities.

Forrester's pseudoscience was embraced uncritically by policymakers in the Nixon administration. Several small cities also adopted his ideas in hopes of encouraging or flattening growth. They didn't question the provenance of Forrester's hundreds of equations, nor did they test whether adjustments to the parameters of those equations would yield different conclusions. Most of these real-world experiments were failures. There was one success, however. Residents of Forrester's hometown in Concord, Massachusetts, approached him with concerns about suburban growth threatening the town's "character" (often used as a racial epithet). He advocated for restrictive zoning laws. By limiting housing, they'd limit growth. This, predictably, drove housing prices sky-high. From 1970 to 1990, Concord's population grew less than 0.05 percent per year, while its home prices appreciated by 11 percent annually. Concord restricted growth and priced out young families in the process. This can hardly be said to reflect a triumph of Forrester's equations; rather, it shows that he pulled from the age-old playbook of using zoning laws to throttle housing supply and exclude specific populations from communities.

Forrester believed that game-based models might one day replace debate, capturing nuance more successfully than language. The issue, he argues, is that "the human mind is not adapted to interpreting how social systems behave." People are decent at linking causes to effects but cannot reason about complex interrelational dynamics. Business consultant William Patterson wrote a warm review of *Urban Dynamics* for the libertarian magazine *Reason*. Patterson clearly understood that Forrester's models were nothing more than models and didn't reflect real data. The variables Forrester identified as important "pressure points" to improve society lacked experimental or factual backing. Nevertheless, Patterson professed hope that "libertarians rather than statists will be the first to discover and utilize such pressure points, to bring about a more free society." Patterson recognized that models could be used to support the conclusions of statists and libertarians alike, depending on what aspects were included. What mattered was who wielded them first, and more forcibly.

SimCity is, by all accounts, an odd game. It has no clear win state. Yet this didn't seem to concern players, who were happy tinkering to accommodate the ever-shifting needs of their growing creations. Players occasionally stumbled on stable equilibrium states, which laid plain the biases hidden in Forrester's equations. Artist Vincent Ocasla "won" the game by creating a city with a stable population of six million. The only catch? The win state was a libertarian nightmare world. It had no public services—no schools, hospitals, parks, or fire stations. His dystopia had nothing but citizens and a concentrated police force populating an endless plain of one bleak city block, copied over and over.

Despite the playfulness of *SimCity*'s design—a monster reminiscent of Godzilla, for instance, randomly attacks densely populated areas—it was often taken seriously as a scientific program. In the 1990 Providence mayoral race, Joseph Braude, a fifteen-year-old freelancer for *The Providence Journal*, arranged a *SimCity* competition between election candidates. Their game performance had a nonnegligible impact on the election outcome: candidate Victoria Lederberg maintained that Braude's negative coverage of her *SimCity* performance cost her the race. In 2002, candidates for the mayor of Warsaw competed against one another in *SimCity*, as did candidates for German parliament in 2013.

Though Wright was never under the illusion that his game was a serious model of reality, many players believed otherwise. Business consultants deluged Wright with requests to design playable models of their industries for training and educational purposes. Wright was initially reluctant—he'd intended *SimCity* to be only a caricature. But he eventually succumbed to pressure and spun off a short-lived business simulation division called Thinking Tools, headed by Maxis cofounder Jeff Braun. Thinking Tools crafted educational games for big-name clients. Chevron, for instance, commissioned an oil-refinery simulation game called *SimRefinery*. Chevron allegedly used the game to train office workers—who may have never entered a physical plant—to better understand the oil refinement process. The game was apparently a success. Much like Kriegsspiel had helped military officers gain a bird's-eye view of the battlefield, *SimRefinery* helped white-collar workers gain a new perspective on their industry.

Thinking Tools nurtured Forrester's hope that simulations might come to replace debate. After Bill Clinton won the 1992 US election on the platform of health-care reform, the Markle Foundation commissioned Thinking Tools to design a hospital-management simulator. Released in 1994, *SimHealth* was played by policymakers and the public alike—including, famously, Clinton's daughter, Chelsea. Maxis marketed *SimHealth* as more than mere entertainment: it was a policy tool and could be used to explore and reason about complex systems. Players assumed the role of a newly elected politician campaigning for health-care reform. They used their finite political currency to promote policies that aligned with the values on which they based their election promises. They could track their policy changes against their stated values using a compass-like indicator that pitted Liberty against Equality and Community against Efficiency—ideals that are, in reality, by no means opposed.

Unlike *SimCity* players, *SimHealth* players could tinker with the model under the hood and adjust hundreds of parameters representing the model's assumptions. Yet tweaking the parameters was not the same as tweaking the models themselves, and the game had a clear ideological bias. Much as in *SimCity*, there wasn't exactly a win state. But *SimHealth*'s values were hard to miss. The game trumpeted a somber funeral march whenever the Canadian-style single-payer socialized medicine plan popped up on the screen. As Keith Schlesinger writes in a review for *Computer Gaming World*, there was one easy way to win: "All you have to do is adopt an extreme libertarian ideology, eliminate all federal health care (including Medicare!), and cut other government services by $100–$300 billion per year." Unfortunately, this could hardly be called a health policy victory, as it left the virtual citizens entirely without health coverage. Even the private insurance companies went bankrupt in the first few months. The game was a flop, and thirty years later, health care remains an intractable issue plaguing American politics.

Whereas *SimRefinery* gave players a new perspective on a complex, though defined, process, the US health-care industry is so complex that *SimHealth* only muddied the waters. Paul Starr, who was a health-care policy adviser to the Clinton administration, dismissed the game entirely. "*SimHealth* contains so much misinformation that no one could possibly understand competing proposals and policies, much less evaluate them, on the basis of the program." He was concerned that people would mistake the game for a legitimate description of reality. He despaired that his daughter, an avid player, accepted the game's libertarian-leaning strategies because that was "just the way the game works." To borrow social theorist Sherry Turkle's phrase, we too easily abdicate authority to simulations. Simulations are ultimately constrained by their creators' assumptions: games are self-contained universes ticking along to preprogrammed logic. They don't necessarily reflect anything fundamental about the world. Models' conclusions are less interesting than the assumptions on which they're based—assumptions that are typically hidden.

Games may still be useful for reimagining society, however. In their book *The Dawn of Everything*, anthropologist David Graeber and archaeologist David Wengrow suggest that playful experimentation was crucial to forming the wildly creative social structures evident across human history. The zone of ritual play, they write, "acted as a site of social experimentation—even, in some ways, as an encyclopedia of social possibilities." More recently, scholars have sought to use game design and game theory to help us invent our way out of rigid social stratifications and worsening social inequalities.

European philosophers once characterized people as pawns in games played by gods, whose decisions were as inscrutable as dice throws. The advent of probability and decision theory transformed people from pawns to players. Previously, leaders augured the outcome of war using random divination. Thanks to games like Kriegsspiel, they came to marshal the future by gaming out optimal strategies. Formerly, the diversity of life was thought to reflect God's intricate planning. The mathematics of games revealed that mindless rules could result in a dizzying variety of forms and strategies. Once, influential political figures had professed that games were a metaphor for social order: each person had their role to play and rules to follow. But as democratic revolutions challenged the notion of divine authority, people hoped to exercise more control over their place in the world.

Most recently, minimax and related optimization techniques convinced technologists and policymakers that society might be ordered and optimized according to rational principles. Games increasingly underpin the architecture of our economic, technological, and social systems. People participating from every corner of the internet move in invisible markets designed to efficiently extract money, attention, and information from users. Our reputations are scored with social media metrics, dating app recommendations, buyer and seller ratings. The age-old metaphor of life as a game paved its way into reality. Through this progression, our belief in our own agency has expanded. Today, we're transitioning from being players to architects of the games we move in. What was once credited to fate has been corralled into the realm of human design.

13
Moral Geometry: Playing Utopia

Utopia is the process of making a better world, the name for one
path history can take, a dynamic, tumultuous, agonizing process,
with no end. Struggle forever.

KIM STANLEY ROBINSON

The game Moksha Patam (today's Snakes and Ladders) is often
credited to the thirteenth-century Marathi saint Dnyaneshwar,
though it may have been invented hundreds of years earlier. It was
designed to teach players about karma and illustrate the precarity of
fate. In it, souls ascend to the heavenly realms, dragged down by vices
(snakes) and buoyed by virtues (ladders). Snakes are labeled with the
names of demons and the vices they represent, and players recount
stories from Hindu mythology as they play. Salman Rushdie writes in
Midnight's Children:

All games have morals; and the game of Snakes and Ladders cap-
tures, as no other activity can hope to do, the eternal truth that for
every ladder you hope to climb, a snake is waiting just around the
corner; and for every snake, a ladder will compensate.

We don't play Snakes and Ladders; the dice do. There is no free will here, only a series of random numbers. Nevertheless, its players experience acutely the benefit of good deeds, and the agonizing karmic setback of sins. The fact that there is no strategy involved is part of what makes the game so appealing to young children. Chance is the great leveler, and a five-year-old is just as likely to roll a series of winning moves as an adult. Nevertheless, it's seductive to ascribe agency to oneself, believing oneself on a hot streak. The seasoned gambler still kisses their dice; they deem themselves "lucky" when they get a roll they'd hoped for.

In the universe of Snakes and Ladders, karma is something that happens to us—perhaps the consequence of a past incarnation's choices yet beyond our current control. But the game's mechanics still incentivize players to aspire to behave ethically. Many games have been co-opted to teach players moral lessons, gaining spiritual associations over centuries of use. The early board game senet taught ancient Egyptians how to navigate their way through the afterlife, democratizing knowledge that was once the purview of elites. Go came to be thought of as a kind of moral meditation: Buddhist monks ranked among the greatest Go players, and the game was believed to promote enlightenment. In ancient Persia, backgammon took on cosmological significance: the pieces were said to represent humans, while dice throws symbolized the wheeling constellations that governed their fates.

Games, in other words, were often used to entrain social norms in their players. In Plato's *Laws*, a character argues that children's games are essential to civic education. They stabilize society by teaching children how to follow rules, a necessary skill for dutiful citizens. As such, he contends, it is essential that children's games never change, lest the players grow up to believe that laws can be changed, too. A game's rules restrict the actions available to its players. Players can take only so many actions on a given turn, which renders them more predictable to their opponents. The same is true of laws and social norms. Games domesticate us, taming the wilderness of social interaction and testing our peers' trustworthiness under the burden of rules. Game rules and social norms reduce the complexity of our tangled social networks and make life a little more navigable.

Of course, children change game rules all the time. The twentieth-century psychologist Jean Piaget regarded this as the foundation of democracy and of healthy moral reasoning. He analyzed the games of children as they rose through a local primary school. As their games evolved, players learned to cooperate in order to compose and enforce new rules. Children came to see rules not as arbitrary decrees imposed by adults but as collectively beneficial for all participants. It therefore made sense to follow them—not out of obligation but by free choice. However, if there was a good reason to change a rule, one could, so long as every other player agreed. Friend groups learned to effectively self-govern, and children grew more autonomous over time as they toyed with and modified their games. This, Piaget argued, is the essence of democracy. Playing children learn to replace the respect of authority with the mutual respect of other players' wills. This realization eventually extends to the moral realm when children come to realize that adopting moral rules—such as not lying or cheating—makes the game better for everyone.

Games were commonly used as tools for instilling social norms and cultivating better citizens. In many cases, this amounted to pasting ethical instructions onto existing game mechanics. But what if games could teach us what norms we should adopt in the first place? Games exemplify the question of how we should best live with one another. The outcome of a player's choice is dependent on, and in turn influences, the choices of other players. Given these dynamics, how do we make fair rules? How do we design a just society given that other players—our fellow citizens—have different preferences, interests, and abilities? Several prominent moral philosophers employed games and game theory to explore these questions.

Game theory and moral philosophy seem strange bedfellows. The rise of game theory as a force in academic thought has not been morally ambiguous: military theorists wielding game theory stripped humanist values from traditional diplomacy, inviting race-to-the-bottom dynamics and a precarious world order. But philosophers and game theorists have also used its techniques to attempt to derive ethics from first principles. A deeper understanding of probability turned the casting of dice from something that simply *was* to something measurable, even, in some sense, predictable. Similarly, whereas morality was something once prescribed by religious texts, put forth by ancient prophets or revered philosophers, Enlightenment thinkers hoped to understand ethics logically and yoke morality to reason rather than divine mandate.

Thomas Hobbes may have been one of the earliest philosophers to offer a game-theoretic account of the social contract. In his *Leviathan*, written after he'd been traumatized by a bloody English Civil War, he seeks to explain why states and governments should exist. How can it make sense to participate in society at all if people are fundamentally selfish and can only be trusted to look out for themselves? The Civil War had made it clear that people were no longer buying the "divine right" of royalty. Instead, Hobbes hoped to convince his countrymen that it was rational to submit to authority. He posits that the original state of nature was a war "of every man against every man." There could be no industry or technological advancement, "no knowledge of the face of the earth; no account of time; no arts; no letters; no society; [. . .] continual fear, and danger of violent death; and the life of man, solitary, poor, nasty, brutish, and short." Hobbes had no magical telescope to spy on the distant human past, of course; this assertion was filtered through his bitter memories of recent war.

How, then, could trust ever arise? What came along to glue society together if God had not planted kings among his creations? A naive dissenter in the text, the Foole, takes up these questions and suggests that it's rational for an individual to break the social covenant if it benefits him—discerning, three hundred years before the invention of game theory, that the solution to a prisoner's dilemma is to defect. In response, Hobbes invokes the Leviathan: a sovereign authority that ensures everyone benefits from social progress. The Leviathan offers people military protection and allows them to benefit from the technological and social niceties that are impossible to maintain in the violent (and imagined) natural state. The Leviathan is the sovereign to which we cede our power, that which prevents us from reneging on our commitments and holds society together for the good of all.

About a century later, the philosopher Jean-Jacques Rousseau revisited the same question: How does social cooperation emerge naturally? He didn't invent a sovereign force to stabilize society, instead positing a new game model for the social contract. This game, the stag hunt, would later become a classic scenario in game theory. Imagine two hunters in a forest who must decide individually, without first discussing it, what game to pursue. They can take down a stag by working together or each bag a hare individually. The stag yields much more meat than the hare, but hunting it requires cooperation. Hares can be successfully hunted alone. If neither hunter knows their partner's choice in advance, it's a risk to choose to cooperate. If one hunter goes after the stag and the other goes after the hare, the first will be left with nothing, while the other will at least get some meat. Unlike in the prisoner's dilemma, defecting doesn't yield a greater reward than cooperating, so the hunters are better off going after the greater goal together. Both cooperation and defection are equilibriums of the stag hunt.

The shape of a game is defined by its rewards. The prisoner's dilemma and stag hunt can be transformed into one another by tinkering with their payoff structures: the prisoner's dilemma becomes a stag hunt when the reward for cooperating is greater than that for defecting. In 1977, game theorist Edna Ullmann-Margalit published *The Emergence of Norms,* in which she argues that social norms emerged to transform prisoner's dilemmas into stag hunts and enforce cooperation. In Margalit's example, two artillerymen stationed at an important mountain pass face the choice of fleeing the enemy or staying and operating their gun together. If both stay, there's some chance that they'll be injured or killed, but they'll manage to arrest the enemy's advance. If both flee, the enemy will take the mountain pass and kill them. If one stays while the other flees, the remaining artillerist will die in battle, buying the other enough time to flee to safety. This is a prisoner's dilemma, so fleeing is the dominant strategy for both participants. Given that fleeing spells death for both soldiers and ensures the enemy's advance, this is clearly suboptimal. Ullmann-Margalit recounts stories of German soldiers in World War I who were chained to their machine guns to avoid this outcome. What if norms are like an invisible social mechanism that chains gunners to their weapons? Ideals like honor and camaraderie bind artillerymen to their duties and ensure the optimal outcome. Acting in accordance with one's values often feels like a reward of its own, and this is exactly how norms might change the reward payoff.

This suggests a potential solution to a great mystery of humanity's past: the rise of large-scale cooperation. In smaller groups with little anonymity, norms can be easily enforced. Since everyone knows everyone else, even minor offenses can be broadcast to the group through gossip and corrected by shaming or shunning the offender. But for larger-scale social experiments like cities, people had to find ways to trust strangers they didn't know and with whom they might never interact again. Several scholars have suggested that the belief in omniscient, punishing gods replaced gossip to serve this function. After all, watched people, as the saying goes, are nice people. People internalized the threat of gossip as conscience and self-policed to avoid displeasing the divine. If the threat of gossip can't chain a gunner to his artillery, perhaps the threat of supernatural punishment is enough.

But this solution isn't feasible in the pluralistic societies of today. Citizens subscribe to diverse cultural and religious beliefs. What's considered rude or amoral in one group might be acceptable in another. Today, up to 15 percent of Americans are agnostic or nonreligious, and the Leviathan is moot: democratically elected leaders don't have divine mandates. Scholars have sought to find a purely logical foundation for the question of morality and how society should be structured. In the mid-twentieth century, philosopher John Rawls began a line of inquiry that culminated in his celebrated 1971 treatise, *A Theory of Justice*. In this work, he posits fairness as the foundational principle, or "first virtue," on which society should be built. Game theory shaped Rawls's thinking on this matter. He imagines members of society locked in something like Nash's bargaining game. In Nash's original scenario, players have to divide a shared good, like a cake. In Rawls's thought experiment, citizens are locked in a bargaining ritual to decide on a fair and moral framework for society. They have only one chance to get it right. All members must unanimously agree on society's governing principles. All are rational and purely self-interested. How, then, can participants ensure that the benefits of society are fairly distributed to everyone involved?

Rawls's solution: they should bargain behind a veil of ignorance, what he calls the "original position." This is a place of innocence: before citizens are born, before they know who they'll be or what family and upbringing they'll have. Behind the veil of ignorance, no one knows their position in society—their ethnicity, skin color, gender, wealth, talents, social status, height, health, and so on. Bargainers, therefore, are incentivized to build a society that leaves no one behind.

Society, Rawls argues, is *just* when organized in this way. People behind the veil of ignorance will agree to share social benefits and harms
equally because they don't yet know whether they'll win the lottery of
life. Rawls's solution leverages the great leveler of chance. No one,
after all, can control what hand they're dealt in life. Traditional societies used dice and lots to make difficult decisions equitable, and so
would moral philosophers. The philosophical question of what attri-
butes the veil of ignorance obscures, exactly, remains a thorny one.

What is the best outcome that the bargainers can hope for? Game
theory offers several solutions. The game theorist John Harsanyi had
been working on the same questions as Rawls, though his work is less
well known for having been expressed mathematically. Rawls and
Harsanyi both agreed that, to ensure a fair solution, citizens should
bargain behind a veil of ignorance, without knowing their future position in society. But they disagreed on what solution the bargainers
should adopt. Von Neumann had proved that, in a two-person zero-
sum game, a player's best strategy is to strive for the maximin, or maximizing the minimum payoff. Rawls believed that this was the best
solution: it allows the worst-off players to do as well as possible, arranging society's benefits such that they most benefit the least advan-
taged. Rawls felt that this solution was most in line with principles of
justice, though he also conceded that there were many possible solutions to his imagined bargain.

Harsanyi argued that maximin was a stringently conservative and
pessimistic solution, as it overindexes on worst-case scenarios. Imagine
a woman, Alice, who lives in New York and hates her job. She recently
received an offer for her dream job in Chicago. If she takes it, there's a
minuscule chance that her flight to Chicago will crash and she'll die.
According to the maximin solution, Alice should stay in her terrible
job in New York because that's better than the worst-case scenario:
taking the job in Chicago but dying on the trip over. A more reason-

able solution would factor in how unlikely it is that she'd die on the plane ride, and Alice would accept her dream job after all. Harsanyi's solution to the bargaining problem, which he dubbed utilitarianism, comes from the Bayesian decision theory he pioneered. Here, players should aim to maximize their expected utility—that is, aim for the highest average payoff for all players, weighing outcomes by their likelihood rather than overindexing on exceedingly rare but dire outcomes.

President Bill Clinton awarded Rawls the National Humanities Medal in 1999, lauding him for demonstrating that "a society in which the most fortunate help the least fortunate is not only a moral society but a logical one." Rawls had made an argument, stripped of all sentimentality, that social goods should be distributed as evenly as possible across society. Unfortunately, his ideas have been embraced in theory but not in practice. Take, instead, another possible bargaining solution: the maximax, which maximizes the best-off players' outcomes. This is often referred to as the "wishful thinking" scenario. Though it's an irrational solution to the bargaining problem, players might still conceivably agree to it if they believed that they might find themselves dealt one of the privileged hands. Bargainers can only hope that it will help the less privileged in a trickle-down fashion, but there's no guarantee that it will. This is exactly what has happened with trickle-down, maximax economic policies, which have been the explicit mandate of conservative politicians for decades. Though the policies have only served to worsen wealth inequality, many voters still seem to believe in them enough to vote, against their own self-interest, for politicians who support them. The American dream of social mobility has become weaponized irrationality, a fantasy leveraged by politicians promising that unregulated capitalism will help everyone rise to the top.

Another problem with Rawls and Harsanyi's account of justice is that it's impossible. At the dawn of human civilization, our ancestors never bargained on a collective definition of fairness. Nor does the veil of ignorance clue us in on how to improve society as it is, given its entrenched injustices. Mathematician Ken Binmore argues that the original position feels so compelling because it describes our innate sense of fairness. However, it doesn't explain it. We can't assume the existence of a certain value when we try to understand the origin of values. Binmore and others have instead used repeated games to model how the concept of fairness might have emerged naturally over time.

In the 1950s, game theorists discovered that some repeated games support an infinite number of stable equilibrium strategies. This became known as the folk theorem, because it was such an obvious outcome that no one bothered formally publishing it. Say that two players must decide how to divvy up a chocolate cake. There's no haggling— each player privately decides on a percentage they'll demand, writes it down, and hands their folded paper to a referee. If these claims add up to more than 100 percent, all players lose, and the referee gets the cake. Otherwise, the players get their asks, and any remaining cake is wasted. Over repeated trials, the players will find a stable equilibrium. Sharing the cake half-and-half seems like a natural solution—but so is one-third and two-thirds, or one-fourth and one-half, and so on, ad infinitum. Any strategy that doesn't add up to more than 100 percent of the cake works. This is the folk theorem: it outlines what solutions are feasible. It doesn't matter if a player is greedy, so long as they're matched with a suitably modest player. Mixed populations of players with different strategies can successfully coexist. However, the fair solution—splitting the cake fifty-fifty—is most efficient.

To understand why that is, philosopher Brian Skyrms imagines, for simplicity, that the players in this chocolate cake game can have only one of three bargaining strategies: demand one-half (let's call these fair players), demand one-third (modest players), or demand two-thirds (greedy players). Though the modest and greedy populations together compose a stable equilibrium, it's an inefficient strategy. Whenever two greedy individuals meet, they'll get no cake. Whenever two modest individuals meet, they'll waste one-third of the cake. The fair strategy is the most efficient one, leading to the least waste and the fittest overall population. Skyrms ran simulations of this repeated game in which he initialized the population with different proportions of these three strategies. Over repeated generations, the population gravitates toward the fair strategy, which wastes the least amount of cake. Fairness can be thought of as a heuristic that helps people choose among infinite feasible strategies the one that benefits the largest number of players. Morals, by another name, are just smart plays.

Because the fair solution is a stable equilibrium, we don't need to invent concepts like religious duty or the Leviathan to steady the social contract. Binmore likens this to modeling the ocean using the right primitives—that is, the most basic elements possible. One could imagine modeling the ocean as water molecules bound by certain forces. Given the right conditions, waves should emerge in the model, but waves shouldn't be included in the model as a primitive. That is, waves shouldn't be an input to the model—they aren't inherent to water—but, rather, should emerge as a natural effect. In the same way, fairness arises over repeated games but isn't assumed as an input.

The insight of the folk theorem—that a repeated game may have, in principle, an infinite number of solutions—fits beautifully with the great diversity of human social structures. There is no one way to structure human society, giving lie to the myth of a monolithic original state. We aren't beholden to rigid biological imperatives. Darwin famously jotted in one of his notebooks that "he who understands the baboon would do more towards metaphysics than Locke." To the contrary, Binmore points out that "we are indeed naked apes, but it doesn't follow that the way to find out about human table manners is to watch baboons dining."

Our instinct for fairness underwrites a universal moral grammar, or, as Rawls puts it, a "kind of moral geometry." Our love of fairness didn't arise from an ancient bargain but may have become instilled in us over the course of evolution. Even other mammals and birds value fairness. In laboratory experiments, macaque monkeys who have been rewarded with a low-value treat, like a cucumber, will throw it away if they see a peer rewarded with fruit. Fairness is an intrinsic value. "Traditionalists are virulently hostile to this suggestion," Binmore writes, "because they think that nothing is written in our genes but the savagery of nature, red in tooth and claw." But the folk theorem demonstrates that there is no barrier to individuals settling on efficient and cooperative strategies. People in many foraging-based societies share high-value foods, like meat, equally, regardless of who made the kill, reasoning from the original position of hunters who do not know, when they leave in the morning, who will return with the best food. People don't mindlessly maximize their expected reward, with fairness being some impossible virtue stifling ruthless social efficiency. It is not a fanciful merit that people struggle to perform but something inbuilt in our characters and essential to our collective well-being.

Models should never be taken at face value—*SimHealth*, for instance, hid partisan values in its premises. The game's outcomes meant nothing outside the play world from which they'd sprung. Games can still help us reason about the space of possible social contracts and discern the borders of possibility—what is, and what is not, feasible for a dynamic. Game theory offers solutions that are consistent with human nature—or, more accurately, the nature of self-interested agents. Since Adam Smith first refashioned selfishness into a virtue, economists have valorized it as the actuating principle driving production and invention. But selfishness is by no means humanity's only merit, and we mustn't focus on it at the expense of others. A person may be roughly approximated as a maximizing agent, seeking to satisfy their own preferences. However, we can't ignore the fact that those preferences usually include fairness and egalitarianism.

Modern thinkers have returned to Plato's hope that games might teach players how to improve themselves and become better citizens. In his 1978 book, *The Grasshopper: Games, Life and Utopia*, the philosopher Bernard Suits argues that, should humans ever achieve a future utopia in which all of our needs are provided for, gameplay could become our primary pursuit. In the past several decades, game designers have taken this a step further. Games will not just be the official pastime of Utopia; they'll also help usher it in. Gamification will transform the drudgery of modern life. Fitness apps promise flawless bodies happily earned; Duolingo avows effortless language acquisition; organization apps make to-do lists palatable to the most ardent procrastinators. Games, they promise, are a technology for perfecting ourselves, body and mind. Eventually, desk jobs will be alchemized into entertaining affairs, education will be made effortless, and even the most tedious tasks will become enjoyable.

Evidently, this has not yet happened. Gamification *can* make everyday tasks more enjoyable. But, as game designer Adrian Hon argues, it can't miraculously make *any* task pleasant. In principle, gamification is a wonderful idea: by leveraging our natural play instincts, we can improve our lives. In practice, it's been hijacked by corporate thought leaders and business interests, applied in the most uninteresting ways imaginable. For starters, most of the mechanics that companies slap on top of their systems aren't actually any fun. Many were originally designed to make games more addictive rather than more pleasurable, developed to enthrall players to game platforms. They've been optimized to extract players' time, money, and attention. These mechanics, dubbed "dark design" by experts, often encourage addictive behaviors. Some games, like *World of Warcraft*, never end. Players can log in anytime, anywhere, and find something engaging happening.

Our instinctive love of gambling informs the design of loot boxes, which players buy for the chance of obtaining rare in-game goods. Other game mechanics scratch players' itch to rise up the leaderboard, rack up achievements, or build collections. Players may grind away for long hours to collect artificially scarce in-game items. Other players pay real money to skip the grind: in 2023, a weapon skin in *Counter-Strike 2* sold for over $400,000.

Gamification mechanics are not just materially extractive: some exploit the deeper human need for purpose and connection. Game designer Jane McGonigal extols the virtues of games: unlike life, she argues, games give players a clear sense of purpose and accomplishment. Games are enjoyable because they make players feel productive. This is also why they're so insidious. Many video games are designed to make players believe that they're improving their skills when they're only getting better equipment as they progress through levels. Games offer players clear-cut tasks and a more easily won sense of accomplishment than they might get in their jobs or home lives. But these simulated achievements can also harm players, who might painstakingly tend to a virtual farm while ignoring their real responsibilities.

The same is true for social connection. Social status is a major component of mental health, and games are often designed to boost players' self-importance. Players are transformed into the game's hero, the richest miner, the village Lothario. Games can become a refuge for players suffering from social anxiety or difficulties in their real-world relationships, making them even more perfect addiction engines. Problematic gaming is highest in players reporting loneliness and psychological distress, though it's unclear if this is a cause, or a result, of gaming addiction. To be fair, many gamers find community through the games they play, but we cannot ignore the fact that games are increasingly designed to capitalize on loneliness. Rehab institutions charge tens of thousands of dollars to treat problematic gamers whose lives and relationships have been damaged.

Gamification has influenced the design of internet platforms, shaping the digital public square. Social networks use mechanisms including likes, reposts, and digital coins to turn posting into a scored game, a dynamic that has famously poisoned the waters of many online forums. Political strategist Steve Bannon studied *World of Warcraft* communities and realized that he could mobilize rootless white males with "monster power" into online troll armies, as happened with Gamergate. The alt-right has since repeatedly adopted this tactic to intimidate its ideological enemies. Gamification can also be used as a form of control: in some Chinese cities, citizens' reputations are scored and their everyday choices shaped by a system of punishments and rewards. Where does the responsibility for moral decision-making sit when people's behaviors are being manufactured in service of corporate and governmental interests?

Gamification is particularly insidious in the workplace. It's premised on the idea that clever design can render otherwise unpleasant tasks enjoyable. It amounts to realigning people's actual preferences with those of their employers, co-opting their reward circuits and intrinsic motivation to quietly instill corporate interests. Gamification, in other words, replaces what people actually want with what corporations want. This has already harmed workers. Amazon warehouse employees and Uber drivers using gamified work platforms are nudged to move just one more package, accept just one more ride. This has been linked to an uptick in worker health issues and accidents: workers are driven too hard, for too long. This might be harmful to not just the workers themselves but also those whom their work affects. Anyone who works with other people—medical workers, HR representatives, social services employees—could be pushed to dehumanize their clients. In the same way that gamified military operations distance soldiers from the wars they wage, gamified work environments might nudge workers into morally ambiguous actions, concealed behind the polished skin of a game.

This likely gives gamification too much credit; it can only go so far. Many jobs that can be easily gamified will more profitably be automated. Any skill that can be scored and optimized in some way is vulnerable to machine replacement. AI advocates have long promised that AI will usher in a postwork society where people can choose to spend their time however they please. The prophets of gamification make a more temperate claim: people will still work, but they'll enjoy it. With the advent of generative models, we're beginning to see a very different dynamic emerge. The roboticist Hans Moravec observed that the skills researchers initially expected to be hard to automate, like chess play, were easier to engineer than "unthinking" skills, like walking. In recent years, generative models have made progress toward automating art and writing, yet we're still years away from automating hard labor. People once imagined that robots would shoulder everyday drudgery, leaving us to create art and poetry. Instead, the reverse has happened.

Another problem with the gamification of everything is that people have become drawn into games they never consented to play, at both individual and societal scales. Take the pickup artist's bible, *The Game*, according to which men "win" at life by sleeping with as many women as possible. If women are the score, they're not players with their own agency and so become unwitting pawns in a game they never agreed to participate in. We see this in many spheres of life. Innocent bank depositors take on the risky bets of their bankers. Retail investors are at the mercy of predatory Wall Street financiers who reckon that the market is zero sum and act in ways that doom it to be true. Fraudsters defend their poor behaviors with the callous counterfeit of common sense: "Don't hate the player, hate the game." If we explicitly model our social and financial systems as games, we must ensure they are games that all members of society agree to play, and ones in which everyone can win.

Suits identified a defining character of games: players willingly accept limitations in their pursuit of goals. A game of golf could be easily "solved" by walking over to the hole and plopping the ball in. But that wouldn't be a game—a game's goal must be fairly earned by obeying the rules and boundaries of the game. In this tradition, the philosopher C. Thi Nguyen frames games as the "art of agency." When playing a game, we agree to adopt some restrictions on our actions. The game specifies what abilities we have and what goals we should pursue. It is no surprise, then, that the study of games began with the hope of laying the foundations for a new understanding of agency, or how actors in the world make choices. Yet game theory provides an impoverished account of agency, given that its players are doomed to relentlessly maximize their rewards. Their choices are already made, latent in the game's structure. In reality, a person's preference for an outcome doesn't mean they want that outcome at any cost, or that they're willing to accept all of its side effects. We usually don't know the consequences of a decision until we make it. Our preferences aren't necessarily preformed: we learn them, come to them through experience, maturity, and moral reflection.

If we hope to build a science of agency, it's crucial to build on a framework in which choices are something freely made and not inevitable. Garrett Hardin used game theory to frame forced sterilization as an inevitable conclusion of cold logic. Military strategists used game theory to argue that countries are trapped in an inexorable arms race, and nuclear weapons proliferated. Many corporate executives talk about AI as though they have no choice but to continue developing it. According to them, the race to develop AI is both inescapable and the greatest moral failure of our time, given that it could result in human extinction. Instead of turning to moral reasoning, executives are content to abdicate responsibility to their game-theoretic models of the world. Ironically, the dangers they highlight—what if a model tasked with making paper clips turns the whole universe into paper clips?—are the direct result of the models being relentless maximizers. AI doomers broadcast the danger of building maximization into our models yet are blinkered to the danger of adopting maximization as a business strategy. Present-day capitalism is already pregnant with the dangers of myopic maximization, with its singular focus on boosting shareholder value.

Games themselves are nevertheless a useful realm for understanding human choice and agency. Games give players a sense of control, which is no small merit. The sense of agency is a critically overlooked part of psychological health. Even a minor impedance to agency—a sticky mouse, say—can be infuriating. The loss of bodily autonomy—becoming paralyzed in an accident—is devastating. Tragedy and comedy alike are rooted in the mismatch of agency when a character's means fail their ambitions. Games afford many players a sense of competence and control and can be immensely therapeutic. A bullied child may take pride in being an expert in-game archer. Yet this doesn't change the reality that the child is being bullied. By the same token, games may be increasingly used to placate abuses of freedom at societal scales, serving as an escapist realm. In an era of eroding freedoms—surveillance systems, dampened social mobility, militarized police forces—games may serve as an opiate for diminished agency. Games may well become our primary pastime in a future utopia, but they can just as easily temper dystopia. Virtual worlds might provide the simulacra of scarce real-world goods, like social status, while a player's real-world freedoms and material well-being are eroded. The more that reality is imperialized by games—the metaverse, the gamification of work, social media—the more vulnerable we'll be to these simulacra.

Perhaps their greatest promise lies in being a technology of empathy. Play cements social bonds. Many mammals deprived of childhood play have emotional problems as adults and fail at relationships compared with nondeprived peers. Games, a universal in human societies, might have emerged as a framework for building mental models of one another. They may have been preserved and elaborated across generations thanks to their benefits in teaching people how to reason about one another. Players must understand the beliefs and preferences of other players to secure the best outcome for themselves. This is called theory of mind: the ability to model the mental states of others (including their emotions, desires, beliefs, and knowledge). Play wrestling arguably evolved to help animals practice rarely visited vertiginous states outside their everyday experience, thereby helping them to better understand their bodies. Similarly, game players imaginatively inhabit new mindsets and characters, and come to better understand themselves. Games are safe social experiments: tools for understanding our opponents, trying on new identities, and testing our relationships.

Games can help players empathize with others. The video game *Salaam* puts players in the shoes of refugees fleeing conflict. Players experience the characters' hardships pseudo-firsthand. Games can be more immersive than book or movie narratives: unlike readers, players often use "I" when describing a game character's experience. There's some science to back this up. Researchers at the University of Wisconsin designed a game to teach empathy to middle schoolers. After two weeks of daily play, some players' brains were measurably changed, particularly in areas known to be involved in empathy. Not all players showed these effects, but those who did scored higher on tests of empathy and emotional regulation at the end of the trial.

Granted, the theory of mind doesn't lead only to empathy; it's also necessary for manipulating others. In Poe's "The Purloined Letter," the police ask a brilliant detective for help recovering a letter being used to blackmail the queen. They've already thoroughly searched the main suspect's house, to no avail. The detective successfully finds the letter, which has been hiding in plain sight. The police had missed it because they expected the letter to be elaborately concealed. Anticipating this, the suspect left it out in the open. To illustrate his reasoning, the detective recounts a story about a child with uncanny psychological insight who made a fortune from his friends playing the hand game odds and evens. The child had worked out a mental model of each of his friends' strategies, then used that to anticipate their next guesses.

Still, the average person is more empathetic than Machiavellian. Games help players see inside one another more clearly, arguably a necessary foundation for ethics. In the same way that Rawls's veil of ignorance encourages people to imagine themselves in anyone's shoes, game theory's selfish maximization takes a different tenor when we extend the boundaries of the self. One hundred years ago, the Alsatian physician-theologian Albert Schweitzer was distraught at the state of the world. He'd recently opened a hospital in present-day Gabon and was doing the relief work he'd long dreamed of doing. However, he found himself dissatisfied and haunted by doubt. He'd always criticized the notion of civilization: he'd witnessed firsthand the damage his "civilized" countrymen wrought in Africa. But what did it really mean, stripped of its racist and colonial connotations? What is true civilization—its apotheosis? Was there some universal ethical foundation that underpinned a just society? He obsessed over the question, filling notebooks with disjointed thoughts, but made no progress after months of deliberation. One evening, near the end of a weeks-long boat trip to reach his hospital, a phrase came to him as the sun was setting behind a herd of hippopotamuses: "reverence for life." This was the principle that he hoped would reify ethics. These three words later won Schweitzer a Nobel Peace Prize.

Reverence for life is the recognition that, as our own being is significant to us, every living thing's existence is significant to it. Schweitzer had struck on the basic selfishness of life and proposed that we not abhor it but revere it. "I am life which wills to live, in the midst of life which wills to live," he writes in his autobiography. Schweitzer was influenced by the Jain philosophy of nonviolence, which sparked in him the "tremendous discovery that ethics know no bounds." He was also likely influenced by the African philosophy that goes by many names in the Bantu language family but is most famously known as Ubuntu: "I am because we are." The philosophy dissolves the boundaries between discrete subjects through the realization of our shared selfishness. Each of us knows we are alive, and some force in us wants to keep living. And this is something that relates us to every other living thing, from kelp to kestrels. This is Rawls's original position, applied to all that lives. It is the living experience of the Golden Rule. Humankind is the conscience of humankind, life the conscience of life. Our integrity is directly reflected by the state of the most vulnerable among us. Ethical action flows with the inexorability of logic once we admit our shared identity as life itself, longing to live. Our charge is only to see this clearly enough so that selfhood is extended to its rightful domain: all of us.

14

Mechanism Design:
Building Games Where Everyone Wins

Ye live not for yourselves; ye cannot live for yourselves; a
thousand fibres connect you with your fellow men, and along
those fibres, as along sympathetic threads, run your actions as
causes, and return to you as effects.

HENRY MELVILL

n 1983, the US FDA cleared cyclosporine as an immune suppressant
for organ transplant patients. In an instant, medical treatments that
were once the purview of science fiction became realizable for hundreds of thousands of critically ill people. The only things missing
were the organs themselves. An idea came to Dr. Barry Jacobs while
he was watching a news report covering the mass death of thousands
of Bangladeshis. He was struck not with the tragedy of their loss but
with "the waste of all those organs lying there." Jacobs's devotion to
capitalism had already outshone his Hippocratic oath: he'd recently
been released from prison, having served time for prescription fraud.
He'd lost his license to practice medicine but would repurpose his
clinical expertise to become a self-styled organ broker.

Jacobs set up a for-profit market to match organ donors (mainly
from developing countries) with wealthy patients needing organs. His

company would harvest the organs "from US citizens and Third World indigents," he told a reporter. "It will be pure, free choice on their part. There will be proper, written informed consent. Since many potential donors can't read, the informed consent conference will be tape-recorded." His company would pay for the donor's flight to a US medical center and take a cut of the final sale price. He planned to exploit Medicare services to cover the cost of the surgery needed to extract the organs. "It's a very lucrative potential business. If the 'haves' want it, they'll have to pay. If the 'have-nots' want it, they'll have to pay, too."

An entirely new market had sprung up overnight. It's not unusual for businesses to move faster than government regulation, but Jacobs's business model inspired such widespread revulsion that it elicited an uncharacteristically rapid response from Congress. In 1984, Senator Al Gore introduced legislation banning the sale of human organs for money. The bill also established a private nonprofit organization to match patients in need with donated organs. Jacobs faded back into obscurity before making national news again in 2000, when he was found guilty of strangling his wife to death, a crime for which he served fifteen years in prison.

Like the fragmented health-care system it sat within, the early US organ-matching process was decentralized and haphazard. Hospitals had no incentives to share patient and donor databases. Organs from deceased donors went to waste for lack of finding suitable matches in time. Perhaps the greatest tragedy was the number of potential live donors turned away. Relatives and friends of patients needing transplants were often willing to donate but found to be immune incompatible. Occasionally, happy accidents would link a pair of donors—as happened with Tia Wimbush and Susan Ellis, two colleagues who discovered, in a chance bathroom chat, that both of their husbands needed kidneys. Neither one had been a match for their own husband, but both happened to be a match for the other's spouse. Their swap was successful, though this sort of serendipity is exceedingly rare. Markets are usually organized around the exchange of money, but this isn't legal for organ exchanges. The incentives were there—donors were willing to donate to loved ones—but the right markets didn't exist yet. Mathematicians would soon remedy this.

Board game designer Reiner Knizia claims that the scoring system is the most important part of designing a new game. It's the score that drives players' behaviors. Similarly, in a subfield of game theory known as mechanism design, or reverse game theory, mathematicians design institutional rules that incentivize players to behave in particular ways. In 2003, game theorists Alvin Roth, Tayfun Sönmez, and Utku Ünver created a "kidney clearinghouse." They developed a mathematical framework that efficiently matches donors to compatible recipients in sometimes complex daisy chains of reciprocal exchange. With the right incentives—like winning a kidney for a loved one in need—people happily donate to strangers. The game theorists put their proposal into practice and established the New England Program for Kidney Exchange in collaboration with fourteen regional medical centers. This and other related efforts, such as the National Kidney Registry, have vastly improved the transplant matching process, connecting patient-donor chains with as many as seventy participants. The matching process has continued to improve, especially since medical workers designed more efficient organ transportation systems linking donors and patients nationwide.

Oskar Morgenstern hoped that game theory would provide economists with a more expressive mathematical language, empowering them to go from describing institutions to inventing new ones. Economists had previously been constrained, much like astronomers, to sit back and observe the experiments that nature had performed on their behalf. But game theory could help them reason about which hypothetical institutions might emerge from rules describing any number of possible games. Economists could reason about entirely new systems with novel rules and player dynamics. They might even rationally design new social structures that would better express shared human values.

Morgenstern was right: the subfield of mechanism design radically changed the nature of economics. In traditional game theory, mathematicians use a game's rules to predict the equilibrium behaviors of players. In mechanism design, economists take a desired behavior and formulate rules that will elicit that behavior. As with game theory, mechanism design operates under the assumption that players are self-interested. It is this same selfishness that makes their behaviors predictable. The hope is to design games or institutions such that, even when players act selfishly, the resulting outcome is globally desirable—like securing kidneys for patients in need without exploiting impoverished people.

Leonid Hurwicz laid the foundations for mechanism design in the 1960s. His ideas, initially overlooked, slowly gained traction, and in 2007, he was jointly awarded the Nobel Prize alongside fellow game theorists Eric Maskin and Roger Myerson. The seed for Hurwicz's original idea came from a long-standing debate among economists on the merits of centrally planned socialism versus free-market capitalism. Which system is best at allocating resources among members of society? In 1920—three years after the socialist revolution in Russia—the economist Ludwig von Mises cast the problem as one of calculation. How do the individuals constituting an economy find an efficient equilibrium? It's a hopelessly tangled problem, Mises contended, for a central planning system to work out. Central planners were doomed to take guesses in the dark. The free market, on the other hand, is something like a transpersonal calculator, performing its distributed computations through the choices of independent people participating in the economy.

In a 1945 article, economist Friedrich Hayek extended this argument, formalizing a new role for information in economics. Individual people know what they want and why they want it. This information flows between individuals and is ultimately processed in collective decisions. Take an auction, where each bid can be thought of as a message. Bids fly by, contributing information about how different agents value an item. The computation ends when the bids, or messages, stop. Each bidder has a private reason for their bid—perhaps the item is a crucial ingredient for their new invention or something the buyer believes will spark a new fashion trend. Such detailed information is invisible to central planners. As a result, they can't price items correctly.

Hayek claimed that because information is dispersed across individuals and not equally apparent to everyone, central planning–based economies cannot work. Hurwicz took this analysis a step further. Information is crucial, and central planners may lack the information necessary to organize a market. But free markets also have a critical flaw: they're often unfair. Participants may cheat, withhold information, or bend the rules to better their position—they're not usually given good reasons to be truthful with one another. In many games, players are rewarded for hiding what they know and what they want. The stoic poker player bluffs to avoid revealing a weak hand. Players of the board game Settlers of Catan hide their strengths and hoard development cards until late in the game to avoid inviting attack. Hurwicz, Maskin, and Myerson created mathematical tools to assess the effect of dishonesty in games and used these tools to invent games wherein everyone can do well by being honest.

Hurwicz introduced the idea of "incentive compatibility." Systems should work with, not against, human self-interest, but they should also be designed in a way that results in socially beneficial outcomes.

The parenting trick of splitting a cake by asking one child to cut it and the other to choose their piece first is a classic example of an incentive-compatible mechanism. It leverages the children's self-interest in the cake to yield as fair a division as possible. Centrally planned economies fail to align with inherent self-interest: people are motivated by profit. But free markets, Hurwicz argued, don't necessarily invite good behaviors. Free markets don't result in equitable or socially beneficial outcomes. Mechanism design can help build systems that satisfy both constraints. However, two principal hurdles come with this approach: players don't necessarily play by the rules, and game designers can't always be trusted.

It's well known that players often benefit from bending or breaking a game's rules. Take an eBay auction, for example. Bidding is necessary for an eBay customer. Yet bidding is the last thing they want to do: it's revealing; it lays bare their desires. New users quickly learn to hide their interest and bid at the last minute. It's easier to swoop in at the end of an auction, hoping that the current highest bidder isn't paying attention, to nab an object and avoid escalating a bidding war. In response, bidders learn to attend to the closing minutes of an auction so that they can combat sniping. A small industry revolves around providing software programs that automatically snipe bids for customers who can't afford to babysit their auctions. Users on eBay can manipulate information in other ways. Because the site has a built-in reputation engine, winning bidders sometimes threaten sellers with poor reviews if they don't drop the items' prices. Buyers might even message other bidders or the sellers to coerce a sale: *This is for my dying grandma. This is for my sick child.* All of this amounts to weaponized information. Game players often hide what they want because information is its own currency. The mechanism designer's goal, then, is to invent games that reward players for being truthful.

Adam Smith recognized the difficulty of designing new systems, given that individuals will act according to their interests and not as social designers would wish them to:

> In the great chessboard of human society, every single piece has a principle of motion of its own, altogether different from that which the legislature might choose to impress upon it. If those two principles coincide and act in the same direction, the game of human society will go on easily and harmoniously, and is very likely to be happy and successful. If they are opposite or different, the game will go on miserably, and the society must be at all times in the highest degree of disorder.

Traditional game theorists took a game's rules as given and assumed that individuals would only use legal moves. "Then there is the true game," Hurwicz writes, "the one like real life, where the strategies and moves people make, some of them contain illegal gains. So you take into account when you write the rules of the game that the players will try to cheat." Designers must consider how players will deviate from the game's rules to benefit themselves, and invent better rules as needed. Most players don't want to participate in games that reward cheaters with winning.

Unfortunately, it's not always easy to predict how players will cheat. On May 13, 1897, Guglielmo Marconi sent the world's first radio message, instantly transforming the ether of frequency space into a concrete public good. Like the internet, radio was a thrilling medium connecting distant strangers. But it was crowded and unregulated—the military, newscasters, businesses, and individuals alike adopted the new technology. Radio's popularity was partly to blame for the *Titanic* disaster in 1912. Rescue efforts were delayed because the ship's call for help, received in Newfoundland, elicited a massive response from amateur radio enthusiasts, who swamped all frequencies swapping rumors. At the time, there was no dedicated emergency broadcast band, so the *Titanic's* distress signal couldn't rise above the noise of secondhand speculation. US Navy Commander Cleland Davis had already been calling for government intervention, warning that "calls of distress from vessels in peril on the sea go unheeded or are drowned out in the etheric bedlam." In 1912, the US government assumed responsibility for allocating portions of the spectrum to responsible broadcasters, reserving space for military use and emergency signals. But by the 1980s, the licensing system was so overtaxed that the FCC turned it into a random lottery system. Speculative investors applied for broadband licenses and occasionally won tidy sums by renting coveted airspace to communications companies. The system had devolved again into chaos.

In the early 1990s, the newly elected Clinton administration was saddled with historic levels of debt after Reaganomics. Government officials searched for change under every proverbial couch cushion, hoping to raise revenue without raising taxes. Cell phones were growing in popularity, but the government had been giving away precious airwaves for free when they could have been charging companies. In 1951, the legal scholar Leo Herzel advocated for the government to lease the airwaves to raise funding, though this wasn't yet technologically feasible. It's a massively complicated allocation problem: international megacorporations and local businesses alike must vie for discrete spectrum chunks in various locations across the nation. The telephone company Pacific Bell hired economists Robert Wilson and Paul Milgrom to advise on the design of the government's new allocation system. Milgrom and Wilson responded by turning it into a game.

The organizers faced a conundrum: How do you price something when you have no idea what it's worth? Wilson and Milgrom's solution was to auction all the licenses simultaneously to anonymous bidders.

Players who wanted a chunk of spectrum worked out among themselves, over successive turns, how much they were willing to pay for it. The nascent internet made the design workable. Buyers placed their bids anonymously, online, in successive rounds. At the end of each round, offers were made public, so bidders could adjust their bids or reconsider what bands to bid on. The auction for a given license ended when no one made new bids, which sometimes took months.

First run in 1994, spectrum auctions have since generated over $200 billion for the US Treasury. Wilson and Milgrom were awarded the 2020 Nobel Prize for their design, and similar auctions have raised revenue for governments across the globe. Though these auctions are often touted as proof of game theory's real-world relevance, mathematician Ariel Rubinstein chafes at this notion. He counters: "To the best of my understanding, they based their recommendations on basic intuitions and human simulations, and not on sophisticated models of game theory." These auctions are shaped as much by common sense as by calculus.

Others argue, on more material grounds, that the auctions shouldn't be celebrated. Because the auctions were anonymous, bidders couldn't collude or make secret arrangements. Theoretically. After the auctions ended, academics found drastic irregularities in their outcomes. It turned out that companies still managed to identify themselves and signal their interests through the bids themselves. Companies punished their bidding competitors by raising retaliatory bids in areas they weren't genuinely interested in. The Texas communications company Mercury PCS signaled to competitors to back off by placing bid amounts where the last three digits indicated the license area code they most wanted. Companies clearly avoided competing for licenses against AT&T, which had a reputation for retaliation. The auctions for licenses that AT&T pursued were relatively empty. It's difficult to prove collusion definitively, but bidders who used signaling tactics paid about 40 percent less than other bidders. The FCC lost billions to businesses weaponizing information. Though mechanism design was created to contend with what Hurwicz called the "true game," it's not always possible to anticipate how players will break the rules.

The spectrum auctions' designers have since introduced changes that make it harder for bidders to collude, but there are other drawbacks to the system. Whether the auctions exceeded or fell short of fundraising expectations depends on what predictions one compares them with. Some pundits breathlessly anticipated that the airwave market would raise half a trillion dollars; others anticipated just a few billion. But Congress's goal was not only to raise money—it also wanted to diversify spectrum ownership among smaller businesses. By this metric, the auctions have been a failure. They've only consolidated power in the hands of a few big corporations. Ninety-four percent of the licenses in the 2014 auction went to just three companies: AT&T, Dish Network, and Verizon. This was because the auctions were primarily designed to raise money, argues economist Eli Noam, rather than to allocate resources more fairly. The auction's designers have tried to improve its design in successive iterations. However, the designers are motivated by their own interests as well. Economist E. Glen Weyl and journalist Stefano Feltri criticized the 2017 spectrum auction, which only raised a third of its projected revenue and failed to lease a third of the available spectrum space. Nor did it diversify spectrum ownership in any way. The designer's consulting firms, on the other hand, made tens of millions of dollars from the effort. Weyl called the 2017 auction a "mass privatization of a public resource."

This brings us to another requirement of successful mechanism design: the designer must be trustworthy. Perhaps the most troubling evidence of designer misconduct can be found in the realm of tech companies. Auctions, no longer the remit of stodgy gallerists or motor-mouthed auctioneers, have become foundational tools for the internet. William Vickrey was the first economist to analyze auctions using the language of games, work for which he was awarded the Nobel Prize in Economics in 1996. Auctions often suffer from what's known as the winner's curse. In a traditional auction, a bidding war can easily lead to the buyer paying an inflated price. This can cause bidders to underbid or spook bidders from participating in auctions altogether. In 1961, Vickrey analyzed a curious auction method used for a century in the sleepy realm of British stamp collectors. He discovered that the design had a remarkable property: it incentivizes bidders to tell the truth.

In what's now known as a second-price, or Vickrey, auction, there's only one round. Buyers submit their bids silently, in sealed envelopes. Whereas the winning bidder of a typical auction pays the price they bid, here winners pay the second-highest bid price, plus one cent. The winner's curse is lifted, and bidders are rewarded for bidding, truthfully, the exact amount they're willing to pay. If they win the auction, they will pay less than they were ready to. If they overbid, they risk paying more than they'd wished. If they underbid, they risk losing to another bidder who offered an amount they would have been willing to spend.

The second-price auction, pioneered in the quiet corner of conflict-averse philatelists, was adopted by Google's advertising platform, AdWords, in 2000. Since its inception, AdWords has generated well over $1 trillion in ad revenue and it accounts for around 35 percent of all money spent on digital advertising. It would be an understatement to credit Google's success to AdWords: today, 80 percent of Google's revenue comes from advertising. AdWords runs about three billion auctions per day. Advertisers want to target their ads to the best-matched audience, and they bid for advertising slots linked to users' search terms and browsing history. The advertiser making the highest bid should theoretically win the slot. Because it's a second-price auction, the winning advertiser pays the amount bid by the second-highest bidder, plus one cent. But charges made in a 2021 antitrust suit filed by seventeen attorneys general led by the State of Texas against Google allege a serious flaw. The complaint contends that, starting in 2013, Google manipulated its ad-space auctions using insider information to favor its clients over those using third–party ad-placement platforms. Since Google runs these auctions, it games the system to benefit the bidders it wants to win—its customers. The engineers code-named this Project Bernanke, after the former chair of the Federal Reserve— apparently in reference to Bernanke's program of printing money, as they expected the project to generate $230 million in revenue in 2013 alone. According to the complaint, the winning advertiser still paid Google the second-highest price. But Google passed only the third-highest bid along to the ad publisher (that is, the site that runs the ad). It pocketed the price difference, then used this excess to inflate bids from advertisers placing their ads with Google Ads over those using competing platforms. Since Google earns a 20 percent cut on ads placed with its service, the complaint alleges it was both inflating the success rate for its clients and potentially pocketing a portion of the price difference. Because publishers were being paid artificially low amounts, Project Bernanke reduced publisher revenue by up to 40 percent. In short, sites were paid less for showing ads on their pages, and Google made more for itself by manipulating its clients' success rates using the information edge it had over its competitors. The attorneys general argued that Google acted as "pitcher, batter, and umpire, all at the same time." A Google spokesperson responded that the suit misrepresented technical decisions the company had made to improve its products as decisions made to squash competition. In a 2019 antitrust conference, Google chief economist Hal Varian admitted that in some instances Google is "on both the buy side and on the sell side," but he could not explain how the company managed both these roles, which he claimed was "too detailed for the audience, and for me." At the time of writing, the complaint joins two related antitrust cases filed against Google by the US Department of Justice, though these three cases have only recently begun oral arguments. Antitrust cases are notoriously difficult to win and will likely take years to settle.

No matter how pristine a mechanism is in theory, people aren't mathematical objects. In the design of real-world markets, one must consider psychology, communication, persuasion, information processing, and enforcement mechanisms. Any information disparity will be exploited. Google built the system, and Google appears to have gamed the system. It has access to information that its players don't, and it uses that to manipulate outcomes in its favor.

In his Nobel lecture, aptly titled "But Who Will Guard the Guardians?", Hurwicz explains why game design alone can't prevent players from cheating. Players gravitate toward a game's Nash equilibrium: strategies from which, by definition, a player doesn't want to deviate, like driving on the right side of the road in the US. In this sense, Nash equilibria are stable and self-enforcing. However, this assumes that players can't or won't cheat in any way. In reality, deviating from an equilibrium strategy using an illegal move may still be profitable. A motorcyclist fleeing police might still drive against traffic to make their escape.

Mechanism design is the science of regulation. Its rules are devised to keep players honest, but it still needs external enforcement to work—that is, some way to punish cheaters. Enforcement mechanisms can look like physical commitment devices, such as one-way traffic spikes that prevent cars from using unauthorized exits. The fact that elected officials serve finite terms is a mechanism that keeps them accountable to their electorate. But these mechanisms can be fragile—

elected leaders have been known to dismantle term limits or invalidate election results.

Commitment devices are important because they make other players more predictable. The philosopher Don Ross gives the example of a pair of poachers aiming to bag a rare antelope. For their plan to work, one must flush the antelope toward their partner, who will kill it and load it onto their truck. Nothing prevents the second poacher from driving away at this point, bagging the entire trophy for himself. So why would they cooperate at all? A commitment device could look like this: The first poacher rigs the truck with an alarm that only he can trigger. If the second poacher drives off without him, he can sound the alarm, causing both to get caught. Strangely, this arrangement is preferable, because it makes both poachers' promises to cooperate credible, allowing them to confidently pull off the crime.

The failures of the collusive spectrum auctions and Google's alleged exploitation of insider information show that players and designers alike will exhibit endless creativity to exploit a system. This includes corrupting regulators. For enforcement to work, Hurwicz offers, society needs an honest, principled judicial system and, above all, whistleblowers—both in private and public spheres—to push back on power abuses. In the past few decades, whistleblowing has become central to the public conversation. Whistleblowers, sometimes demeaned as "leakers," have revealed stunning government and corporate abuses, from NSA surveillance overreach to war crimes. Concerned citizens have pushed back on police violence in the US by recording footage of police interactions on their cell phones. The EU adopted mandatory corporate whistleblowing protections in 2019, and some corporate protections exist in the US. But protections for federal workers and private citizens are notoriously patchy. Whistleblowing is often a dangerous and thankless task, though regulators can sometimes incentivize it with rewards. A whistleblower who presents evidence of their employer's financial fraud to the SEC, for example, is handsomely remunerated. Some cities are flirting with programs that offer cash to citizens who submit evidence of parking violations, effectively turning the public into agents in a distributed panopticon. But not all whistleblowing is incentivized. People who raise nonfinancial concerns, like sexual and racial discrimination, often suffer for their service. Daniel Ellsberg, the game theorist who leaked the Pentagon Papers detailing US war crimes in Asia, miraculously avoided prison but faced a lifetime of villainization, as though stained by the wrongs he had revealed. Nor is whistleblowing necessarily effective: after Ellsberg's revelation, the Vietnam War continued for another decade. At minimum, whistleblower protections are necessary. No matter how rationally designed a system is, power abuses will arise. Game theory can be used to identify the limitations of what's possible, but it cannot sanitize the real world from the need for morals.

Mechanism design increasingly permeates our lives. It suffuses the internet. Tech companies employ a cadre of economists who design and tweak the rules of their service's game. LinkedIn has formalized the labor market, helping employers find and hire qualified job seekers and establish a measure of reputation substantiated by trusted connections. Rideshare apps match riders with (nominally vetted) drivers. Dating apps match prospective partners who employ different strategies to stand out amid a jungle of options. Subtle tweaks to a platform's rules can elicit different participant behaviors. The dating app Bumble solves a common problem—female users being swamped by low-effort messages—by requiring female users to message first. Video game companies employ economists to design in-game economies—that is, the rules governing the production and flow of resources in the game—which shape players' experiences.

Designers now face the challenge of including ethical guardrails and reintroducing values like trust into their products and markets. As author Cory Doctorow points out, software protocols have become a battleground over what values are included or codified into our realities, dictating the games we're forced to play as end users. In Adam Smith's original conception of capitalism, people are not motivated by selfishness alone. They also value their reputations and statuses, which encourages them to act honestly. Reputation is like a commitment device: to protect their good name, a player is less likely to defect in a prisoner's dilemma.

Yet trust is increasingly hard to come by. With the advent of the internet, the public sphere has grown privatized, and software now mediates our experience of formerly organic interactions, turning human connection into a marketplace. For instance, dating app users are notorious for treating their matches as expendable, sometimes dehumanizing them for personal convenience. Behaviors that were once verboten are more common within largely anonymous environments where poor behavior is unlikely to affect a user's reputation in other spheres. Women, LGBTQIA+ users, and people of color face disproportionate abuse, sexual harassment, and stalking on internet platforms. Some users are casually racist on dating apps in ways they wouldn't dare be on professional social networks.

This is by no means the first time that people have sought ways of formalizing shared virtues into their social systems. Democracy, a core value of many modern states, is endlessly imperiled by the fragility of its principles. If we are to honor it, we must discover ways to ensure that people's preferences are expressed and fairly represented. Political theorists have adopted game theory as a tool to imagine new voting strategies and election reforms. A famous difficulty in collective polling—whether to survey voter preferences in democratic elections or to measure customer satisfaction in the market—is capturing people's true preferences. Game theorist Kenneth Arrow won the Nobel Prize in Economics for discovering what's now known as Arrow's impossibility theorem. He made measuring preferences a more precise affair. Say that a voting population must choose among three or more options. Is there a way to discover a "true" ranking of these options by polling people? Given a small number of reasonable assumptions, Arrow proved that this is impossible. For example, suppose that voters are choosing between three candidates—A, B, and C—and they submit their preferences in ranked order. It's conceivable that two-thirds of the voters prefer A over B, two-thirds prefer B over C, and two-thirds prefer C over A—making it impossible to rank the candidates in aggregate.

There is no ideal voting system, and no mathematics can solve this. But that doesn't mean that all voting systems are hopelessly flawed; they simply have trade-offs. Every voting system has technological and theoretical weaknesses and strengths. Mathematicians can use the tools of game theory to analyze different voting structures, determine the power of different voters and voter blocs, and more. The US, the UK, and Canada all use plurality voting, a system in which the candidate who polls highest wins, regardless of whether they receive a majority of votes. Experts widely criticize this method. Unlike majority voting, plurality voting fails to faithfully represent voters' interests. It inevitably leads to two-party systems, which stifle innovation. Several countries, including Ireland and Australia, have adopted an alternative voting method called ranked-choice voting, which results in fairer, more representative outcomes.

The same question at the heart of democracy—how do we best measure people's preferences?—is also asked by companies hoping to meet customers' expectations, COOs polling worker morale, teachers seeking feedback on their classes, and buyers seeking trustworthy ratings of products and services. Most voting systems take the rich nuance of opinion and flatten it into a binary signal: yes or no. But some allow voters to dial up or down the volume of their preferences. Anyone who has ever filled out a survey or left a customer review is familiar with the Likert scale, a one-to-five ranking poll that has been used to gauge preferences, pain levels, satisfaction, and so on for nearly a century. And anyone who has tried to parse meaning from a sea of reviews lodged by people with vastly differing expectations for a product knows how difficult it is to aggregate meaning across customers' conflicting criteria.

In 1961, economists proposed a pricing mechanism that has recently been recast and popularized as quadratic voting. In quadratic voting, voters can express their preferences at a finer level of granularity. Instead of being given a single vote for different issues, all voters are given an equal number of tokens to vote with as they wish. They can vote for an option multiple times if they strongly prefer it. However, each successive vote costs quadratically more tokens. One vote costs one token, a second vote costs four tokens, a third costs nine tokens, and so on. Economist E. Glen Weyl, one of the popularizers of quadratic voting, established the nonprofit RadicalxChange to advocate its use in real-world governance, be it online, local, corporate, or otherwise. In 2019, the Democratic caucus of the Colorado House of Representatives used quadratic voting to choose which spending bills to prioritize. Taiwan's government runs an online platform, Join, which uses quadratic voting to directly poll citizens' concerns. Citizens are given ninety-nine vote-tokens, which they use to express how they'd like to prioritize political issues. Weyl hopes that people will experiment with quadratic voting and incorporate it into everyday democratic processes in their workplaces and communities.

The blockchain is another, much more controversial technology whose designers sought to formalize a value in game-theoretic terms—in this case, trust. Bitcoin's pseudonymous inventor, Satoshi Nakamoto, devised a proof-of-principle solution to a problem posed by opaque financial institutions. The financial crisis of 2008 underscored the fragility of the US's centralized banking system and revealed how financiers had concealed corrupt practices in a morass of exotic financial derivatives. Banks are meant to act as a trusted third party that records financial transactions, but Nakamoto imagined a decentralized alternative. Bitcoin operates as a public financial ledger shared collectively by network participants. But for this system to work, it can't let just anyone add transactions. There must be a way to discern legitimate transactions from fraudulent ones. Nakamoto used mechanism design to build a ledger by consensus.

Usually, a bank would serve as the arbiter of real transactions. Here, miners compete for the right to write the next block of transactions on the public ledger, work for which they're rewarded a newly minted Bitcoin. Only transactions that get enough votes from the network are considered valid and added to the ledger. Network participants must collectively agree on a ground truth. Bitcoin is not, however, immune to bad actors: if a majority colluded, they could cheat the system and rewrite the ledger in their favor. Bitcoin doesn't do away with trust entirely, but it does make trust explicit. It's a probabilistic trust: the ledger is trustworthy so long as most miners don't collude to tamper with it.

Given its practical limitations—it's currently far too slow to serve as a global payment processing system—Bitcoin is most interesting as an experiment in mechanism design, enforced through code. The second-largest name in the blockchain space, Ethereum, grew out of a similar commitment to decentralization. Its inventor, Vitalik Buterin, had been an avid *World of Warcraft* player until his favorite Warlock spell was nerfed in 2010, when he was fifteen. When a game company finds an imbalance in a game, it may weaken (nerf) or strengthen (buff) the imbalanced element—like replacing an automatic rifle with a Nerf gun, or vice versa. Buterin joked, "I cried myself to sleep, and on that day I realized what horrors centralized services can bring." He quit the game in disgust and discovered Bitcoin, impelled by his distaste for centralization. Bitcoin had already inspired hundreds of copycat projects, but Buterin realized that these were often focused on hyperspecific problems. He invented a more generalizable alternative, Ethereum. Ethereum's smart contracts can be custom programmed for different purposes, similar to how computers have come to replace music players, phones, and calculators. Because they can be programmed to enforce transactions, smart contracts ensure that market participants uphold their agreements. This acts like a commitment device, warping game payoffs to enforce cooperation, much like the antelope poachers' alarm system or the chaining of soldiers to their machine guns.

Blockchain technologies are adversarial by design. They were created on the assumption that network users will act selfishly. But given that mechanism designers can't always anticipate how users will game the system, smart contracts are difficult to code. Unfortunately for many users, blockchains are also immutable—there is no "undo" button. As such, the ecosystem is rife with multimillion-dollar hacks capitalizing on minor vulnerabilities in the code, effectively using the blockchain's rules against itself. There is a "dark forest" of bots combing through posted Ethereum transactions, looking for exploitable weaknesses. Hackers steal billions of dollars' worth of cryptocurrencies every year. Some say these are necessary growing pains. Our understanding of this technology grows with—and is hardened against—every new vulnerability revealed by its exploiters. To these believers, the end goal—a trustworthy, transparent system robust to manipulation by small, powerful groups—is worth fighting for.

As we've seen, a major drawback of mechanism design is that the designer must be trustworthy. But designers are not easily held accountable. Those who have the power to tamper with the rules of a

game in their favor will likely do so. Rubinstein likens this to a playground full of children:

> How did we choose which game to play? We understood that chess
> is for deep thinkers, that Scrabble is for the fluent of tongue, and
> basketball is intended for the tall. Everyone liked a different game.
> We agreed on which game to play by balancing our conflicting de-
> sires, and because we wanted to stay friends. I think of the choice of
> economic policy in the same way as I think of the childhood choice
> of which game to play.

Different games suit different players, and sometimes a player uni-
laterally changes the game's rules—usually the strongest kid on the
playground. In a game-theoretic model, players are bound by a game's
rules; in reality, they often bend, break, or tamper with them. In theory, players have a limited, circumscribed set of options or actions to
choose from. In the real world, a sufficiently creative, wealthy, or powerful player may invent any number of possibilities for themselves,
including some that exist outside the game. Mechanism designers
attempt to temper this, but it's ultimately inescapable. And, increasingly, powerful corporate interests play the role of both mechanism
designer and Rubinstein's strong kid, changing the rules to benefit
themselves.

Play, historian Johan Huizinga argues in his classic book *Homo
Ludens*, is how humans innovate, from new tools to new social contracts. People were never handed rules for getting along. They've
needed to build and sometimes break social rules to explore new orga-
nizing principles. The playground offers such a laboratory. Take, for
example, the preponderance of cultures with a Carnival season, loos-
ening social restrictions to allow for new interactions across the social

hierarchy. Play is generative, a form of collective ideation. Huizinga sees games as a foundational cultural technology: "Civilization arises and unfolds in and as play." Playful individuals and societies may fare better because they're more adept at generating and adopting new behaviors.

Mechanism design is the formalization of Huizinga's instinct, promising to help us conceive fairer social systems. It's essential, however, that we take up this work in the spirit of play. Designing better systems will require an iterative approach that adapts over time to change along with the world. Play has served as a crucible of culture and innovation; it's at the heart of design itself. Physicist Michael Nielsen writes:

> At its deepest, design is about inventing entirely new types of objects and of actions. It's the type of thinking that results in someone conjuring up the rules of chess; or inventing topological quantum computing; or in developing the polymerase chain reaction. In each case we take the rules of reality and find latent within them some other very different set of rules, a set of rules for a new reality, one that generates beautiful patterns of its own.

Design is what happens when we uncover rules latent in the world and use these to define the logic of a new, separate system. The inventors of chess abstracted the different capabilities of army divisions to invent the rules of a game that's entertained us for centuries. We can't change reality, but we can tinker with designed systems to encourage more desirable behaviors. There's benefit to creating rule-based environments: realms whose laws are known and whose consequences are predictable. A deeper understanding of games empowered researchers to design novel marketplaces, educational products, and voting systems.

But games are inevitably an oversimplification of reality. As games have continued to imperialize our everyday lives, they've come to replace traditional values and relationships with rigid and unrealistic models. They've ushered in the gamification of everything, resulting in addictive systems where all worth is metered by monetary exchange.

No model or mechanism, in the end, can correct the imperfections of reality. Games temporarily suspend the rules of daily life, allowing players to enter alternate moral universes where they're rewarded for swindling their sibling, lying to their partner, or shooting their friend. Humans owe their success to systems—of beliefs, laws, institutions, values—that enforce and empower wide-scale cooperation. Mechanism design is a thinking aid that can help devise systems that make cooperation easier and more frictionless. It's a way to implement positive changes that are more resilient to gaming. It offers muted utopias: best possible worlds but never perfection. Thomas More, after all, coined *utopia* to mean "no place," a clever homonym to *eutopia*, "a good place." For truly effective social reform, we can use the tools of game theory to understand why we're in the equilibrium we're in and to map out realistic alternatives. We should be willing to iterate on entirely new game structures, not just tinker with the rules of games we're already in. Politicians, policymakers, and technocrats must resist sweeping claims about inevitable tragedies of human nature based on abstract models. The fact that people play so many different games demonstrates the fundamental flexibility of our value system. As ever, we must remain vigilant against mistaking models of reality for reality.

Epilogue

Man plays only when he is in the full sense of the word a man,
and *he is only wholly Man when he is playing.*

FRIEDRICH SCHILLER

n his story "On Exactitude in Science," Jorge Luis Borges imagines an empire whose geographers have perfected the art of cartography. They create a map so detailed that it takes up as much space as the land itself. The map waxes and wanes as the empire grows and shrinks through military campaigns and defeats. As successive generations come to realize the map's uselessness, it is left to dissolve in tatters across the countryside. In the early 1980s, the philosopher Jean Baudrillard argued that the opposite is true of modern life. We increasingly live in our maps of reality instead of in reality itself. The territory has been replaced with towering layers of abstractions: sovereign nations, states, counties, personal properties, deeds, mortgages, mortgage-backed securities, mortgage-backed credit default swaps, and so on. It is, he writes, "the territory whose shreds are slowly rotting across the map."

This is especially true for the games that underpin so many of our technologies. Games are a particularly dangerous fiction because they shape how their players act and make choices. Games are like maps that remake the world in their own image. The rules and rewards of a game dictate how its players behave. A game can replace people's actual preferences with those rewarded by its scoring function. They can even be used to endow unconscious computer programs with the illusion of intention. These programs amount to mathematical functions yet can appear to talk, desire, and act in ways specified by their programmers, simply because they were designed to maximize some desired quantity. People, unlike these mathematical functions, are much more than reward optimizers. Yet we aren't immune to a game's influence. Game theory is not a very good model of people, but it's good enough to be trouble.

Game theory, it was promised, would help us solve pressing collective problems like nuclear proliferation, climate change, and pandemic outbreaks. It fed into the fantasy that the mind—a product of biology—works in orderly ways, amenable to clean mathematical analysis. Channeled through the right filter, we might perfect human nature and clarify our minds. But humans are not merely players, acting to maximize preprogrammed preferences. What economists have long classified as biases in human behaviors are not flaws in our brains' computation. In many cases, they are the computation itself. Humans are imperfect learners—a reality that economists must seriously contend with. The notion of bias is particularly insidious because it suggests that people can't reliably make choices in their own interest. As a result, economists, technologists, and policymakers sometimes deem themselves justified in substituting their own.

In fact, game theory promised to automate decisions altogether: we'd simply follow the choices dictated by its synthetic rationality. This was a seductive promise in the era of nuclear diplomacy, when decisions made by a few people might have undone the world. It would have been no one's fault if their models had failed: they'd simply obeyed the inexorable logic of the situation. They'd acted rationally.

But what might the world look like if more people in power weren't so afraid of taking responsibility for their choices? What if decisions weren't concentrated in the hands of so few? What if more systems operated using true democratic practices, rather than replacing people's choices with models of those choices?

Game theory has been used to launder arbitrary, sometimes harmful ideologies, often mistakenly held up as objective truth. Neither games nor any models are truly value neutral. Some famous game-theoretic notions won mindshare by way of clever marketing: Game theory's agents are "rational," not "greedy." Economic surplus is re-labeled "social welfare." It's the inadequacies of humans, not their economic models, that give rise to "cognitive biases." The tenets of game theory have been co-opted by corporations to justify anti-union and anti-regulation policies, to manipulate public opinion, and to rationalize widespread exploitation. As it's commonly practiced, game theory is based on the assumption that players are single-mindedly maximizing some measurable reward. For decades, economists contended that growing the GDP would cure all social ills. But by many sociological measures—longevity, suicide statistics, and addiction rates—Americans have not fared well. Relentless maximization is the logic of a cancerous tumor, not of health. The adoption of game theory in real-world systems has led to a quiet abandonment of ethics. In such systems, anything that can't be quantified becomes invisible instead.

If game theory were a legitimate account of agency, its agents wouldn't be helpless to their desires. They could change their minds and discover new goals. But game theory is not an account of human behavior. It has not, and cannot, reveal what people truly want. Game theory's focus on individual decision-making obscures larger social, political, and historical factors that shape human behaviors. Play emerged over evolution as a form of learning and a way to strengthen social bonds. For mathematical simplicity, game theorists often exclude our social nature and the fact that we learn what to want from one another and from our circumstances. We need to be aware of the ways that the games encoded in our social and technological systems warp our preferences, manufacture our desires, contaminate our attitudes toward one another, and ultimately influence our life choices.

Economists make much of incentives, crediting them as the engine of innovation, failing to recognize that play is itself a powerful intrinsic drive. People are motivated by many things besides money: the joy of discovery, production, security, a stable family life, the company of their colleagues. Real people are not as constrained as model players. Within the universe of chess, a player's agency is limited to their legal moves. But true agency draws on free play. A real agent can invent new actions, new rules, even entirely new games. They may learn, midgame, that they want something different than what they initially believed and change their strategy accordingly. People change. Even games change—modern chess is very different from its ancient analogue. We need to embrace models that better reflect this reality.

Though game theory is not an ideal microscope for dissecting human behaviors, researchers might have better luck using games themselves. In Orson Scott Card's novel *Ender's Game*, officers probe their students' psychology using the immersive Mind Game. The program's advanced AI responds to unique player choices, offering customized outcomes and character interactions depending on their responses. Using automatically generated scenarios, it encourages the students to make intellectual and emotional breakthroughs. AI-driven game companies are now hoping to make this fictional technology real.

Games will be a transformational learning technology: they're a wonderful model of how we naturally learn. Players make choices and experience their consequences, acquiring knowledge firsthand instead of theoretically. Designers of games and novel user interfaces—like Bret Victor's Dynamicland—have incorporated how people instinctively learn to create revolutionary forms of media. Rather than demanding that people mold themselves to awkward cognitive systems—clunky programming languages, for instance—we can create programming interfaces that more naturally fit with our cognition and encourage new modes of thinking.

Video games are better than they've ever been. Some evangelists promise that they'll even anesthetize the drudgery of the modern office and temper the loneliness plaguing so many people. Others worry that generative AI will usher in even more addictive and immersive experiences. Games may cease being collective spaces and instead become custom fit to each player, fractured along eight billion individual fault lines. Games may be generated on the fly, adaptively reshaping themselves in dialogue with players' interests and moods. Text programs will spin off targeted narratives. Generative art programs will invent environments exquisitely tailored to each user's aesthetic taste. Human designers will simply polish the mirrors lining the interiors of these solipsism engines.

These games will be two-way mirrors, however. Games can be used to generate massive amounts of human behavioral data, a prized feature for today's data-hungry technologists. Programs may be used to infer players' internal states through their decisions and physiological measures. Whereas scientists once used noisy data like click-through rates and buying patterns to measure people's preferences, emerging VR and wearable technologies offer unprecedented surveillance. Companies will collect data on eye movements, pulse rate, skin conductance, facial expressions, posture, and locomotion. Metrics that were once the gold standard of top psychology labs can now be collected at scale by private corporations. Almost every aspect of the user's sensory input and physiological output will be analyzed. The explicit goal of surveillance capitalism is to keep as much measurable content within the system as possible. By the same token, the explicit aim of the metaverse is to turn more of life into a game. Potential choices will be offered by, and limited to exist within, the system itself. The map of the territory will grow to match the territory. But it still won't *be* the territory. Games are self-contained and based on rules in a way that reality is not. We should be wary of trading our autonomy for entertainment.

For thousands of years, people have aspired to purify their thinking through gameplay. Scholars repurposed games to derive the principles of intelligence, morality, trade, and military strategy. There's certainly power in recognizing what games we've built for ourselves, and how we might consciously design better ones. But by creating systems built on greedy maximizing functions, we've invented new monsters to slay. The problem is not that a paper-clip-maximizing AI will arise in the future and turn the universe into paper clips. The maximizers are already here. Any consequences too subtle to measure—environmental costs, civic discord, troubled diplomatic relations—are simply omitted from the score. Rapacious business interests cannibalize useful companies for profit, and workers and consumers are relentlessly squeezed. Military strategists marginalize morality in favor of credible bluster. Race supremacists mistake evolution for optimization. Social media platforms maximize attention from their users. These functions are already embedded in our social systems and cultural beliefs. They are programmed to win at any cost.

Acknowledgments

'd like to thank and celebrate the many people whose support and wisdom helped shape this book. I'm especially grateful to my editors, Courtney Young and Laura Stickney, for their brilliant guidance, to Catalina Trigo and Fahad Al-Amoudi for their expert shepherding through the publication process, to Lauren Morgan Whitticom for her immensely insightful edits and corrections, and to the many kind folks at Riverhead and Penguin Random House, without whom this book wouldn't be possible. Thanks to my agent, Will Francis, who championed this book from the start—it has been such a pleasure working with you. I'm obliged to Ayshan Aliyeva, who fact-checked this book with enormous skill. A huge thank-you to my early readers, in particular John and Jon Chatlos, Aaron Koralek, Albert Kao, and Lowry Kirkby. I'm profoundly indebted to the wisdom of professors Anita Desai and Bharati Mukherjee, who taught me so much about writing. I'm also grateful to the hundreds of scholars, designers, and activists whose work I've referenced—I've learned an enormous amount from you all. Special thanks to all those who took the time to speak with me; I hope I've faithfully represented your ideas. This book was generously supported by the Alfred P. Sloan Foundation Grant for Public Understanding of Science and Technology.

I'm endlessly grateful to my love, John Chatlos—it's been a joy to explore these ideas in conversation with you. Thank you to my family—Robert, Martha, Leo, Laura, Caitlin, Maira, Maggie—particularly for encouraging my early love of science, and to my wonderfully welcoming new family—Jon, Jo El, Suzannah, Tom, Jane, and Harrison. I am, as always, honored by the unflagging support of my friends—Maria Acosta, Aaron Koralek, Andrea Gomez, Raquel Martins, Nicole Hampton, Steve Lynch, Lowry Kirkby, Hassana Oyibo, Dot Amesbury, Ivana Orsolic, Jarvia Foxter, Jacques Bothma, Albert Kao, Ali Bevers, Vishal Maini, and many others. I'm forever humbled by your brilliance and humor. This book would not exist without the moral support of strangers on the internet who post pictures of their pets. And, finally, I am thankful for Ruth Barnett and people like her, whose quiet integrity may go uncelebrated but not unnoticed.

Notes

Chapter 1: The Play of Creation

3 *"Lila* is the play": Burroughs, *Selections from the Gospel of Sri Ramakrishna,* 130.

3 "fruit of Arithmetic": Moyer, *Philosophers' Game,* 39.

3 a "banquet proper": Alan of Lille, *Anticlaudianus,* 379–80.

8 "they saw the evil": Herodotus, *Herodotus,* 43.

9 "Fun is just another": Koster, *Theory of Fun for Game Design,* 46.

10 Play is practice: Spinka, Newberry, and Bekoff, "Mammalian Play," 141–68; and Groos, *Play of Animals,* 141–68.

11 Limestone boards carved: Rollefson, "Neolithic Game Board from 'Ain Ghazal," 1.

11 "a little digital drug": Anderson, "Just One More Game . . ." 28.

12 "We can imagine that": Feynman, Leighton, and Sands, *Feynman Lectures on Physics.*

14 "no theory changes": Skinner, *Beyond Freedom and Dignity,* 215.

15 "The tools we use": Dijkstra, *Selected Writings on Computing,* 129.

Chapter 2: How Heaven Works

16 "To learn is a natural": Aristotle, *Aristotle's Treatise on Poetry,* 108.

16 A British doctor recalled: Hoffman and Vilensky, "Encephalitis Lethargica," 2247.

17 pulled out all her teeth: Hoffman and Vilensky, 2247.

17 A seventeen-year-old boy: Sridam and Phanthumchinda, "Encephalitis Lethargica–like Illness," 1521.

17 It was first synthesized: Marsden, "Dopamine, S136–44"; and Stein and Thiel, "History of Therapeutic Aerosols," 27.

17 devoutly took six drops: Bhargava and Kant, "Health File of Mahatma Gandhi," 18.

17 Sarpagandha was later "discovered": Roy, "Global Pharma and Local Science," 277.

17 the Swedish researcher: Carlsson, Lindqvist, and Magnusson, "3,4-Dihydroxy-phenylalanine and 5-Hydroxytryptophan as Reserpine Antagonists," 1200.

18 **Katharine Montagu's 1957 discovery:** Montagu, "Catechol Compounds in Rat Tissues and in Brains of Different Animals," 244; and Yeragani et al., "Arvid Carlsson, and the Story of Dopamine," 87.

18 **a Viennese neurologist:** Hornykiewicz, "Dopamine Miracle," 502.

18 **"involuntary tremulous motion":** Parkinson, "An Essay on the Shaking Palsy," 223.

19 **"extinct volcanoes" erupting:** Sacks, *Awakenings*, xxv.

19 **Scientists have since discovered:** Yamamoto and Vernier, "Evolution of Dopamine Systems in Chordates," 21; Cottrell, "Occurrence of Dopamine and Noradrenaline in the Nervous Tissue of Some Invertebrate Species," 63; Kindt et al., "Dopamine Mediates Context-Dependent Modulation of Sensory Plasticity in C. *elegans*," 662; and Kass-Simon and Pierobon, "Cnidarian Chemical Neurotransmission," 14–20.

20 **"no pretensions whatever":** Lovelace, "On Babbage's Analytical Engine," 722.

20 **"the objection says":** Turing, "Computing Machinery and Intelligence," 450.

20 **"Instead of trying to":** Turing, 456.

21 **he writes, "is the computational":** McCarthy, "What Is Artificial Intelligence?"

21 **Both Turing and Michie:** Pearce, "Donald Michie, 83, Theorist of Artificial Intelligence, Dies."

21 **"something unspecified but romantic":** Srinivasan, *Donald Michie*, 11.

21 **Thanks to his work:** Andresen, "Donald Michie," 82.

22 **"I resolved to make":** Boden, "Obituary," 765.

22 **He built a learning system:** Michie, "Experiments on the Mechanization of Game-Learning Part I," 232.

23 **"an extremely idiotic character":** Michie, 234.

23 **understand "animal stupidity":** Thorndike, "Animal Intelligence," 3.

23 **named this the "law of effect":** Thorndike, *The Elements of Psychology*, 166.

24 **chess was to the study:** McCarthy, "Chess as the Drosophila of AI," 227.

24 **"the use of chess":** van den Herik, "Computer Chess Today and Tomorrow," 13.

25 **Schultz and his colleagues implanted:** Schultz, Apicella, and Ljungberg, "Responses of Monkey Dopamine Neurons to Reward and Conditioned Stimuli during Successive Steps of Learning a Delayed Response Task," 900.

26 **"Throughout my career":** Kumar and Yeragani, "Penfield—A Great Explorer of Psyche-Soma-Neuroscience," 276.

26 **a postdoc named James:** Olds and Milner, "Positive Reinforcement Produced by Electrical Stimulation," 419.

26 **game of hot and cold:** Olds, "Physiological Mechanisms of Reward," 84.

27 **"Evidently all the desirable":** Asimov, *The Human Brain*, 188.

27 **Stimulating this area:** Olds, "Brain Stimulation and the Motivation of Behavior," 401.

28 **Surprise indicates that something:** Foster and Keane, "The Role of Surprise in Learning," 75–85.

29 **"The real question":** Skinner, *Contingencies of Reinforcement*, 265.

29 **designed "psychocivilized" society:** Delgado, *Physical Control of the Mind*, 23.

29 **"It is a mistake":** Skinner, "'I Have Been Misunderstood,'" 63–65.

29 **Heath announced that:** Moan and Heath, "Septal Stimulation for the Initiation of Heterosexual Behavior in a Homosexual Male," 27–30.

30 "Surely someone else would": Sutton, "Episode 11—Richard Sutton," 13:25.
30 In his junior year: Klopf, *Brain Function and Adaptive Systems.*
31 "a guess from a guess": Sutton and Barto, *Reinforcement Learning,* 124.
32 Backgammon's branching factor: Berliner, "Computer Backgammon," 66.
32 Tesauro named his program: Tesauro, "Temporal Difference Learning and TD-Gammon," 58.
33 "in some cases to major": Tesauro, 63.
33 Sejnowski reached out: Sejnowski, "Dopamine Made You Do It," 259–60.
33 This elegant framework: Montague, Dayan, and Sejnowski, "A Framework for Mesencephalic Dopamine Systems Based on Predictive Hebbian Learning," 1936–47; and Schultz, Dayan, and Montague, "A Neural Substrate of Prediction and Reward," 1593–99.
33 "about the pursuit of happiness": FORA.tv, "Dopamine Jackpot!," 1:06.
34 It's responsible for reward-seeking: Barron, Søvik, and Cornish, "Roles of Dopamine and Related Compounds," 163.
34 "Learning and planning": Sutton, "Fourteen Declarative Principles of Experience-Oriented Intelligence."
35 Perhaps hallucinations result: Brisch et al., "Role of Dopamine in Schizophrenia from a Neurobiological and Evolutionary Perspective," 47; Maia and Frank, "Integrative Perspective on the Role of Dopamine in Schizophrenia," 52–66; and Millard et al., "Prediction-Error Hypothesis of Schizophrenia," 628–40.
35 argues psychologist Alison Gopnik: Gopnik, *Philosophical Baby,* 129.
39 the "Aha!" moment of insight: Tik et al., "Ultra-high-field FMRI Insights on Insight," 3241–52.
39 "People who enjoy solving": Schell, *Art of Game Design,* 35.
39 caliph of Baghdad: Shenk, *Immortal Game,* 1–3.
39 King Canute of Denmark: Larson, *Canute the Great,* 222–23.
39 people still die: Kuperczko et al., "Sudden Gamer Death," 824.
39 young men have disproportionately: Aguiar et al., "Leisure Luxuries and the Labor Supply of Young Men," 35–38.
39 Consumers spend more money: Dudley, "Business of Gaming."
39 Problematic gamblers make up: Degenhardt et al., "Global Epidemiology and Burden of Opioid Dependence," 1320–33; Calado and Griffiths, "Problem Gambling Worldwide," 592–613; and Stevens et al., "Global Prevalence of Gaming Disorder," 553–68.

Chapter 3: Dice Playing God

41 "All chance, direction": Pope, *Works of Alexander Pope,* 103.
41 "a notoriously ineffective method": Moore, "Divination—A New Perspective," 69.
42 "is almost wholly": Speck, *Naskapi,* 128.
42 In the game Rochambeau: Fréchet, "Commentary on the Three Notes of Emile Borel," 121.
43 nervous system of many moths: Miller and Surlykke, "How Some Insects Detect and Avoid Being Eaten by Bats," 570–81; and Yager, "Predator Detection and Evasion by Flying Insects," 201–7.

43 Many mammals use: Moore et al., "Unpredictability of Escape Trajectory Explains Predator Evasion Ability," 440.

43 birds aren't always able: Mills, Taylor, and Hemelrijk, "Sexual Size Dimorphism, Prey Morphology and Catch Success in Relation to Flight Mechanics in the Peregrine Falcon," e01979; and Greggor et al., "Pre-Release Training, Predator Interactions and Evidence for Persistence of Anti-predator Behavior," e01658.

43 The Dogon read: Peek, *African Divination Systems*, 140.

43 The Dalai Lama: Maurer, Rossi, and Scheuermann, *Glimpses of Tibetan Divination*, x, xix.

43 "a real act of faith": Sommer, "Gambling with God," 271.

44 In countless cultures across: Culin, "American Indian Games (1902)," 58–64; Dotson, Cook, and Lu, *Dice and Gods on the Silk Road*, 1–21; and Perry-Gal, Stern, and Erlich, "Gaming and Divination in the Hellenistic Levant," 65–79.

44 seven-thousand-year-old human settlements: Watson, *Archaeological Ethnography in Western Iran*, 199.

44 "As alluring as a draught": Bhide, "Compulsive Gambling in Ancient Indian Texts," 294–95.

44 The flow of valuable: DeBoer, "Of Dice and Women," 216.

45 "winning from yourself": Goodwin, *Social Organization of the Western Apache*, 375.

45 "like an imbecile": Steinmetz, *Gaming Table*, 165.

45 "almost any flat surface": Lanciani, "Gambling and Cheating in Ancient Rome," 7.

45 "when Constantine abandoned Rome": Steinmetz, *Gaming Table*, 168.

46 "I do not know": France, "Gambling Impulse," 376.

47 The Russian writer Fyodor: George, "From the Gambler Within," 226–31.

47 "Games of chance came into vogue": Steinmetz, *Gaming Table*, 187–88.

47 "found cards and dice": Dusaulx, *De la passion du jeu, depuis les temps anciens jusqu'a nos jours*, 94.

47 "At the death of Louis XIV": Steinmetz, *Gaming Table*, 99.

47 To bring the situation: Barnhart, "Gambling in Revolutionary Paris," 151–66.

48 suggests physicist Shmuel: Sambursky, "On the Possible and the Probable in Ancient Greece," 47.

48 "some hidden principle": Plutarch, *De Stoicorum repugnantiis*, 454.

49 "a habit of mind": David, *Games, Gods and Gambling*, 26.

49 "because noe 2 thynges": Recorde, *Whetstone of Witte*.

50 "with such ease men die": Craig, "Hamlet's Book," 17.

50 "kids, lambs, hares, rabbits": Cardano, *Book of My Life*, 53.

50 "counterpoise to an insane": Cardano, 282.

50 "immoderately given to gambling": Cardano, 54.

53 "Let us weigh": Pascal, *Thoughts, Letters, and Minor Works*, 85.

54 "The reader will soon": David, *Games, Gods and Gambling*, 115.

54 "Although in games depending": Huygens, *Christiani Hugenii libellus de ratiociniis in ludo aleae*, 2.

55 Laplace would later put it: Laplace, *Philosophical Essay on Probabilities*, 196.

55 "only the expression": Laplace, 3.

57 "never a moment ceasing": Cummings, *Complete Poems*, 756.

59 "An immediate risk": Lipkowitz, "Physicians' Dilemma in the 18th-Century French Smallpox Debate," 2329.

59 "The true foundation": David, *Games, Gods and Gambling*, 103.

60 Society itself contains: Quetelet, *Treatise on Man and the Development of His Faculties*, 6.

60 "There are three kinds": Twain, *Autobiography of Mark Twain*, 228.

61 "blinded by bell curves": Taleb, *Fooled by Randomness*, 241.

63 "Maybe is addictive": FORA.tv, "Dopamine Jackpot!," 2:04.

64 This explains the counterintuitive: Anselme and Robinson, "What Motivates Gambling Behavior?," 182.

64 Schizophrenic patients with: Speers and Bilkey, "Maladaptive Explore/Exploit Trade-Offs in Schizophrenia," 341–54.

64 side effect of dopamine-enhancing: Heiden, Heinz, and Romanczuk-Seiferth, "Pathological Gambling in Parkinson's Disease," 67–72.

65 people's brains respond to new information: Kobayashi and Hsu, "Common Neural Code for Reward and Information Value," 13061–66.

65 Gamblers report euphoric feelings: Schüll, *Addiction by Design*, 18.

65 Game designer Greg Costikyan: Costikyan, *Uncertainty in Games*, 2.

66 Because surprising information: Foster and Keane, "The Role of Surprise in Learning," 75–87.

67 If we assess the: MacAskill, *What We Owe the Future*.

Chapter 4: Kriegsspiel, the Science of War

71 "Play is the beginning": Dorsey, *Why We Behave Like Human Beings*, 355.

72 "the most close representation": Leibniz, "A Note on Certain Games," 23.

72 "I strongly approve": Leibniz, *Die philosophischen Schriften von Gottfried Wilhelm Leibniz*, 304.

73 They used dice to: Wintjes, "Europe's Earliest Kriegsspiel?," 22.

74 It became a favorite: von Hilgers, *War Games*, 44.

74 He made the game: Wintjes, "When a Spiel Is Not a Game," 8.

75 "It's not a game": Caffrey, "Toward a History-Based Doctrine for Wargaming," 35.

76 "nerveless body" of local states: Hamilton, Madison, and Jay, *Federalist Papers*, 126.

76 "can take part immediately": Leeson, "The Origins of Kriegsspiel."

77 "The ability to quickly": Vego, "German War Gaming," 110.

81 The science-fiction writer H. G. Wells: Wells, *Little Wars*.

82 "If we do not end war": Menzies, *Things to Come*.

82 Anthropologist Michelle Scalise: Kniffin and Sugiyama, "Toward a Natural History of Team Sports," 211–18.

83 Some scholars have argued: Gregory, "Chunkey, Cahokia, and Indigenous Conflict Resolution," 38–54.

Chapter 5: Rational Fools

85 "But it is clear": Rilke, *Letters to a Young Poet*, 62.

86 "What he really yearned for": Hannak, *Emanuel Lasker*, 7.

86 "What is struggle and victory?": Lasker, *Struggle*, 12.

87 "Humanity stands before": Lasker, *Lasker's Manual of Chess*, 249.

87 "That wars should appear": Lasker, *Lasker's Manual of Chess*, 250.

88 "The unconscious motive": Jones, "The Problem of Paul Morphy."

88 The German mathematician Ernst: Zermelo, "Über eine Anwendung der Mengenlehre auf die Theorie des Schachspiels," 501.

88 What constitutes a winning position: Weintraub, *Toward a History of Game Theory*, 733.

88 "a code that determines": Borel, "On Games That Involve Chance and the Skill of the Players."

89 "Whenever I talked with": Wigner, "Two Kinds of Reality," 261.

90 "Johnny was the only": Halmos, "Legend of John von Neumann," 386.

91 "intuitively as immediate experience": Hilbert, "New Grounding in Mathematics," 202.

91 "Mathematics is no longer": Weyl, "Current Epistemological Situation in Mathematics," 123.

91 "The fundamental idea": Hilbert, "Foundation of Mathematics," 239.

92 "a priori do not": Vonneuman, "John von Neumann As Seen By His Brother," 3.

93 "It is easy to picture": von Neumann, "On the Theory of Games of Strategy," 21.

94 "the principal problem": von Neumann, 13.

94 "It makes no difference": von Neumann, 23.

94 "to realize Hilbert's goal": Ulam, "John von Neumann 1903–1957," 11–12.

95 "My personal opinion": von Neumann, *Neumann Compendium*, 623.

95 "I know myself how": von Neumann, 623.

96 "some arrested emotional development": Seeger, "Von Neumann, Jewish Catholic," 235.

96 "I feel the opposite": Leonard, *Von Neumann, Morgenstern, and the Creation of Game Theory*, 185.

97 "the whole affair is": Leonard, 204.

97 "In the state of the": Leonard, 204.

98 "Unproductive as it is": Leonard, 211.

98 "Which 'winning' coalition": von Neumann, "Theory of Games I (General Foundations)," 13.

98 "I cannot accept that": Leonard, *Von Neumann, Morgenstern, and the Creation of Game Theory*, 214.

99 "I believe that these": Leonard, 211.

100 "hardly deserve[s] the names": Schotter, *Selected Economic Writings of Oskar Morgenstern*, 180–81.

100 "The foresight of whom?": Morgenstern, "Perfect Foresight and Economic Equilibrium," 171.

100 "Robinson Crusoe's experiences": Marx, *Capital*, 88.

101 "We hope to establish": von Neumann and Morgenstern, *Theory of Games and Economic Behavior*, 2.

102 "It is just as foolish": Wigner, *Collected Works of Eugene Paul Wigner*, 130.

103 In 1937, the economist: Samuelson, "A Note on Measurement of Utility," 155–61.

103 "The modern [utility] theory": Binmore, *Natural Justice*, 98.

104 "gave me great satisfaction": Leonard, "From Parlor Games to Social Science," 753.

104 As the Nobel Prize–winning economist: Sen, "Rational Fools."

105 "orders of society": von Neumann and Morgenstern, *Theory of Games and Economic Behavior*, 42.
105 "One of the ironies": Mirowski, "What Were von Neumann and Morgenstern Trying to Accomplish?," 143.
105 "a dynamic theory would": von Neumann and Morgenstern, *Theory of Games and Economic Behavior*, 44.
105 "Johnny says we should": Mirowski, "What Were von Neumann and Morgenstern Trying to Accomplish?," 143.
106 Players in a two-person zero-sum: von Neumann and Morgenstern, *Theory of Games and Economic Behavior*, 128.
107 "the contrast of attitudes": Shubik, "Game Theory at Princeton, 1949–1955," 153.
107 "Mr. Nash is nineteen": Duffin, "Nash, John Forbes, Jr., 1950," 1948.
108 "The state of equilibrium": Cournot, *Researches into the Mathematical Principles of the Theory of Wealth*, 81.
108 "That's just a fixed-point": Nasar, *A Beautiful Mind*, 94.
110 The ascendancy of free-market: Oreskes and Conway, *The Big Myth*.

Chapter 6: The Clothes Have No Emperor

112 "What makes you so": Rota, *Indiscrete Thoughts*, 58.
112 In many game-theoretic: Birch and Billman, "Preschool Children's Food Sharing with Friends and Acquaintances," 387–95; Brownell, Svetlova, and Nichols, "To Share or Not to Share," 117–30; Warneken and Tomasello, "Emergence of Contingent Reciprocity in Young Children," 338–50; and Samek et al., "Development of Social Comparisons and Sharing Behavior across 12 Countries."
113 "Mathematical theorems are tautologies": Binmore, *Game Theory and the Social Contract*, 95–96.
113 "operations research man": Shubik, "Game Theory and Operations Research," 4.
115 "probability theory is nothing": Laplace, *Philosophical Essay on Probabilities*, 196.
115 "The stinker," the habitually: De Herdt, "Cooperation and Fairness," 188–89.
115 "the *E. coli* of": Axelrod, *Complexity of Cooperation*, xi.
116 many of its classic: O'Grady, "Fraudulent Data Raise Questions about Superstar Honesty Researcher"; and Lee, "Weird Research-Misconduct Scandal about Dishonesty Just Got Weirder."
116 At a 1966 conference: Archibald, *Strategic Interaction and Conflict*, 147.
117 "despite its beauty": Gintis, *Bounds of Reason*, xvi.
120 Players behave differently: Liberman, Samuels, and Ross, "Name of the Game," 1175–85.
120 Gintis posits that norms: Gintis, "Five Principles for the Unification of the Behavioral Sciences," 17.
120 "It is the *logic*": Ross, "Game Theory."
120 It's perfectly possible: Gintis, *Bounds of Reason*, 82.
121 "is therefore the final step": Binmore, *Natural Justice*, 184.
123 "I am not sure": von Neumann and Morgenstern, *Theory of Games and Economic Behavior*, 635.
124 "members of the Victims": Rubinstein, *Economic Fables*, 115.
124 "doubts whose very existence": Hesse, *Glass Bead Game*, 95.

125 **The more juice:** Schultz, Carelli, and Wightman, "Phasic Dopamine Signals," 147–54; and Schultz, "Dopamine Reward Prediction Error Coding," 23–32.

126 **They register relative differences:** Burke et al., "Dopamine Receptor-Specific Contributions to the Computation of Value," 1415–24.

127 **A person's susceptibility to:** Norton and Sommers, "Whites See Racism as a Zero-Sum Game That They Are Now Losing," 215–18; Kuchynka et al., "Zero-Sum Thinking and the Masculinity Contest," 529–50; and Różycka-Tran et al., "Belief in a Zero-Sum Game and Subjective Well-Being across 35 Countries," 3575–84.

128 **details what zero-sum thinking:** McGhee, *Sum of Us.*

129 **"As a rational being":** Hardin, "Tragedy of the Commons," 1244.

129 **"freedom to breed":** Hardin, 1246.

130 **"inherent logic of the commons":** Hardin, 1244.

130 **Governments have since used his:** Radkau, *Nature and Power,* 72.

130 **"my position is that":** Hardin, "Living in a World of Limits."

130 **"the children of improvident":** Hardin, "Tragedy of the Commons," 1246.

130 **his philosophy "lifeboat ethics":** Hardin, "Living in a World of Limits."

131 **"formal approaches, fieldwork":** Ostrom and Ostrom, "Rethinking Institutional Analysis."

131 **"fallible learners who seek":** Ostrom, "A Behavioral Approach to the Rational Choice Theory of Collective Action," 9.

131 **"We are neither trapped":** Ostrom, 16.

131 **criticized for being "too complex":** Nijhuis, "Miracle of the Commons."

131 **"simply reject it":** Ostrom, "Beyond Markets and States," 665.

131 **"I still get asked":** Nijhuis, "Miracle of the Commons."

131 **"A resource arrangement that":** Fennell, "Ostrom's Law," 9.

132 **Hardin's ideas have been:** Partelow et al., "Privatizing the Commons," 749.

134 **it is as if decision-makers:** Sugden, "Community of Advantage," 420.

Chapter 7: A Map That Warps the Territory

137 **"The chief reason warfare":** Arendt, *Crises of the Republic,* 107.

138 **"Perhaps my factories will":** Tägil, "Alfred Nobel's Thoughts about War and Peace."

138 **"I begin to believe":** Baker and Clarke, *Letters of Wilkie Collins,* 344.

138 **"impossible alike from":** Bloch, *Future of War,* xi.

139 **"the public conscience became":** Strauss, *Men and Decisions,* 351.

139 **"oldest and most blood-drenched":** Poundstone, *Prisoner's Dilemma,* 80.

139 **"obscene interest in computation":** von Neumann to Veblen, May 21, 1943.

140 **"that all his involvements":** Glimm, Impagliazzo, and Singer, *Legacy of John von Neumann,* 2.

140 **"It will not be sufficient":** von Neumann, "Defense in Atomic War," 1090.

140 **"where the intellect has":** Rapoport, "Use and Misuse of Game Theory," 114.

141 **"violently anti-communist":** von Neumann, "Nomination of John von Neumann to Be a Member of the United States Atomic Energy Commission."

141 **"there is no cure":** von Neumann, "Can We Survive Technology?," 672.

141 **"With the Russians":** Blair, "Passing of a Great Mind," 97.

142 **Historian S. M. Amadae points out:** Amadae, *Prisoners of Reason,* 21.

143 **"he never really felt":** Bott, *Mathematics Related to Physics,* 270.

Notes

17

143 "uninhibited by ethical considerations": Seeger, "Von Neumann, Jewish Catholic," 236.
143 She surmised that: Dyson, *Turing's Cathedral*, 272.
143 "which he had never": Blair, "Passing of a Great Mind," 104.
143 "von Neumann suffered more": Poundstone, *Prisoner's Dilemma*, 193.
144 "some people confess guilt": Ulam, *Adventures of a Mathematician*, 224.
144 "He felt with steadily": Vonneuman, "John von Neumann As Seen By His Brother," 64.
145 In his 1960 book: Schelling, *The Strategy of Conflict*.
145 the "vicious diplomacy": Schelling, *Arms and Influence* (1966), 2.
146 "Let us assume that": Kahn, *Thinking about the Unthinkable in the 1980s*, 59.
147 "Your generals talk of": Schelling, *Arms and Influence* (2020), 39.
147 "balance of fear": Sherwin, "One Step from Nuclear War."
148 "the tactic of deliberately": Schelling, *Strategy of Conflict*, 200.
148 "I call it the Madman Theory": Haldeman and DiMona, *Ends of Power*, 83.
149 "the unprecedented risks": Keeny and Panofsky, "Mad versus Nuts," 287.
149 "There is no such thing": Perry, "Herbert York Dies at 87."
150 "a preprogrammed president": York, *Race to Oblivion*, 232.
150 "public policy could itself": Eisenhower, "President Dwight D. Eisenhower's Farewell Address."
150 "We can learn our jobs": Reeves, *President Kennedy*, 25.
151 "He gloried in the graphs": Brodie, "Why Were We So (Strategically) Wrong?," 156.
151 At RAND, he'd already: Poundstone, *Prisoner's Dilemma*, 168.
151 "The future is uncertain": Hitch, "Uncertainties in Operations Research," 443.
152 "fell flat on their faces": Brodie, "Why Were We So (Strategically) Wrong?," 151.
152 "This goddamned bombing": Milne, *America's Rasputin*, 216.
152 "a feeling is widely": Karnow, *Vietnam*, 520.
153 "What are we to make": Davis, *Lawrence and Oppenheimer*, 330.
153 "I can go into my office": *First Use of Nuclear Weapons: Preserving Responsible Control, Hearings before the Subcommittee on International Security and Scientific Affairs*, 94th Cong. 218 (1976).
154 "learn from mistakes": Badham, *WarGames*.
155 "Would you prefer a warm": Kahn, *Essential Herman Kahn*, 22.
155 "How selfish soever man": Smith, *Theory of Moral Sentiments*, 1.
156 "I think it's a very tempting": Rubinstein, "The Best Books on Game Theory."
156 "It teaches us what": Rapoport, "Use and Misuse of Game Theory," 118.
157 "WELCOME TO THE BATTLEFIELD": Warner, *RobotWar*.
158 "The fact ARTUµ was": Roper, "AI Just Controlled a Military Plane for the First Time Ever."
159 "military-entertainment complex": Sterling, *Zeitgeist*, 122.
159 This distancing reduces: Nelson, Wilson, and Kurina, "Post-Traumatic Stress Disorder among U.S. Army Drone Operators," 562–70.
160 By virtually experiencing acts: Hopp, Parrott, and Wang, "Use of Military-Themed First-Person Shooter Games and Militarism," 192–99.
161 "We might even call": Nguyen, "Competition as Cooperation," 123.
163 "decisions make us": Saramago, *All the Names*, 29.

Chapter 8: Chess, the *Drosophila* of Intelligence

167 "In chess, a game of pure thought": Zweig, *Collected Novellas of Stefan Zweig*, 131–32.

167 "away from politics": Zweig, *Married to Stefan Zweig*, 260.

167 "miles and miles away": Prochnik, *Impossible Exile*, 3.

168 "There was nothing to": Zweig, *Collected Novellas of Stefan Zweig*, 120–21.

169 "He looks dead": Prochnik, *Impossible Exile*, 348.

170 "bittersweet creature against which": Miller, *Greek Lyric*, 62.

170 "fills us with wants": Plato, *Plato*, 57.

170 "desire is the very": Spinoza, *Essential Spinoza*, 93.

170 "Paleontologists often say": Erwin, "Call to the Custodians of Deep Time," 282.

172 "The white stones": Leibo et al., "Autocurricula and the Emergence of Innovation from Social Interaction," 1.

172 These bacteria-like life forms: Daubin and Szöllösi, "Horizontal Gene Transfer and the History of Life," a018036.

175 the Machiavellian hypothesis: Barrett and Henzi, "Social Nature of Primate Cognition," 1865.

175 British anthropologist Robin Dunbar: Dunbar, "Neocortex Size as a Constraint on Group Size in Primates," 469–73; and Dunbar, "Social Brain Hypothesis and Its Implications for Social Evolution," 562–72.

175 An individual from a social wasp: Lihoreau, Latty, and Chittka, "Exploration of the Social Brain Hypothesis in Insects," 442.

176 animals that mate monogamously: Shultz and Dunbar, "Evolution of the Social Brain," 2429–36.

176 Scholars have discovered that: Page, *Diversity Bonus*, 15–23; Berner et al., "Dota 2 with Large-Scale Deep Reinforcement Learning," 3; and Zahavy et al., "Diversifying AI," 1–38.

Chapter 9: The End of Evolution

177 "What is life?": Rosen, *Life Itself*, 11.

177 William Paley published: Paley, *Natural Theology*, 7.

178 Darwin saw in: Darwin, *Autobiography of Charles Darwin*, 120.

178 "the law of higgledy-piggledy": Darwin, *Life and Letters of Charles Darwin*, 37.

179 to more "external accidents": Arbuthnot, "Argument for Divine Providence," 189.

180 "the common good differeth": Hobbes, *Leviathan* (2012), 94–95.

180 "but I now see": Darwin, *Origin of Species*, 399.

180 "Natural selection plus time": Keynes and Harrison, *Evolutionary Studies*, 212.

182 "the study of the agencies": Galton, *Memories of My Life*, 321.

184 "In the world of our model": Hamilton, "Genetical Evolution of Social Behaviour," 16.

185 "may go haywire": Harman, *Price of Altruism*, 53.

186 Hamilton called this: Hamilton, "Extraordinary Sex Ratios," 484–86.

190 "we are born selfish": Dawkins, *Selfish Gene*, ix.

191 "The book of life": Noble, *Music of Life*, 10.

192 Using game-theoretic modeling: Zhang et al., "Integrating Evolutionary Dynamics into Treatment of Metastatic Castrate-Resistant Prostate Cancer"; and Staňková et al., "Optimizing Cancer Treatment Using Game Theory."

192 "what we are seeking": Queller, "Gene's Eye View, the Gouldian Knot, Fisherian Swords and the Causes of Selection."

193 "It was so simple": Harman, *Price of Altruism*, 167.

193 "quite a miracle": Harman, 178.

193 "A model that unifies": Price, "Nature of Selection," 389.

193 "Selection has been studied": Price, 389.

194 process as "evolutionary epistemology": Popper and Eccles, *Self and Its Brain*, 540.

195 "destroyed the poetry of": Haydon, *Life, Letters and Table Talk of Benjamin Robert Haydon*, 201.

195 "Give to everyone": Harman, *Price of Altruism*, 217.

195 "My name is George": Harman, 218.

195 "I have less faith": Harman, 212.

196 "Dr. Price is better": Maynard Smith, *On Evolution*, vii–viii.

196 "a completed work of art": Harman, *Price of Altruism*, 280.

197 "In this process": Hamilton, *Narrow Roads of Gene Land*, 323.

197 "powerful constraint on the geometry": Frank, "Price Equation Program."

198 "still in progress": Fisher, *Collected Papers of R. A. Fisher*, 620.

Chapter 10: Nous ex Machina

200 "In a riddle whose answer": Borges, *Collected Fictions*, 126.

200 computers could one day: Shannon, "Programming a Computer for Playing Chess," 256.

201 "admits of being put": Toole, *Ada, the Enchantress of Numbers*, 118.

202 "A game between two": Shannon, "Programming a Computer for Playing Chess," 6.

202 "the level of play": Shannon, 11.

203 "These must be balanced": Shannon, 14.

203 "based on an introspective": Turing, "Digital Computers Applied to Games," 293.

204 a prominent definition of intelligence: McCarthy, "What Is Artificial Intelligence?"

205 "I cannot do that": Schaeffer, *One Jump Ahead*, 99.

205 "It seems reasonable to assume": Samuel, "Programming Computers to Play Games," 165.

206 "the competitive and commercial": McCarthy, "What Is Artificial Intelligence?"

207 "I haven't seen a checker": St. George, "Crown Him."

208 "Chess is like looking": Gardner, "Mathematical Games," 22.

208 "I have a better programmer": Hartston, "Can God Beat the Computer?"

208 "I don't want to let my programmer down": Hartston, "Particularly Susceptible to Draughts."

208 "You're going to regret that!": Schaeffer, *One Jump Ahead*, 188.

208 "What's this guy talking about?": Schaeffer, 188–90.

208 "I had a dream": Schaeffer, 399.

209 "Checkers has been solved": Schaeffer, 540.

210 "America wants you to": Johnson, *White King and Red Queen*, 174.

210 **"was treated by people"**: Kasparov, "Garry Kasparov on Chess and Politics in Soviet Russia."

211 **"simple picture of a large"**: Schaeffer, *One Jump Ahead*, 49.

211 **"Maybe if you threw"**: Thompson, Oral History of Ken Thompson, 6.

211 **Feng-hsiung Hsu studied**: Hsu, *Behind Deep Blue*, 30.

213 **"In certain kinds of positions"**: Hsu, 176.

213 **"I used to think chess"**: Weber, "Mean Chess-Playing Computer Tears at the
Meaning of Thought."

213 **"It reminds me of"**: Hsu, *Behind Deep Blue*, 241.

214 **"I'm not afraid to admit"**: Krauthammer, "Be Afraid."

214 **"The match was never"**: Hsu, *Behind Deep Blue*, 264.

214 **"No effort was devoted"**: Krauthammer, "Be Afraid."

215 **"In the past few years"**: Roeder, *Seven Games*, 80.

216 **"I knew that if"**: Bushnell, "E29."

217 **"any drunk in any bar"**: Jorgensen, *Encyclopedia of Consumer Brands*, 21.

217 **"Whether you were talking"**: Edwards, "Robots, Pizza, and Sensory Overload."

219 **computer scientist Michael Bowling**: Bellemare et al., "Arcade Learning Environment," 253–79.

Chapter 11: Cogito Ergo Zero Sum

222 **"Experience, the mother"**: de Cervantes, *Don Quixote*, 126.

223 **"He must be an important figure"**: Ulam, *Adventures of a Mathematician*, 142.

223 **"almost with suspicion or wonder"**: Ulam, 142.

223 **"the physics having something"**: Ulam, 143.

223 **"At the back of the book"**: Ulam, 144.

223 **"a pure mathematician who"**: Ulam, 6.

224 **"require more multiplications than"**: Ulam, 213.

224 **"if not exactly humiliating"**: Ulam, 197.

225 **"sunny place for shady people"**: Maugham, *Strictly Personal*, 156.

225 **The algorithm was implemented**: Metropolis et al., "Equation of State Calculations by Fast Computing Machines," 1087–92.

226 **"conversation, debate or negotiation"**: An and Ormerod, *Relentless*, v.

227 **"How would nature play Go?"**: Brügmann, "Monte Carlo Go," 1.

228 **"I had no idea"**: British Go Association, "History of Go-Playing Programs."

228 **"The key to a successful"**: Sutton, "Verification, The Key to AI."

230 **"We thought it wasn't"**: D'Agostino, "DeepMind's David Silver on Games,
Beauty, and AI's Potential to Avert Human-Made Disasters."

230 **"found solutions that made"**: D'Agostino.

230 **"not a human move"**: Metz, "Sadness and Beauty of Watching Google's AI
Play Go."

231 **"I thought AlphaGo"**: Google DeepMind, *AlphaGo—The Movie*

231 **"an entity that cannot"**: Yoo, "Go Master Lee Says He Quits Unable to Win over
AI Go Players."

231 **"AlphaGo look like the"**: Google DeepMind, *AlphaGo—The Movie*

232 **"as soon as it works"**: Vardi, "Artificial Intelligence," 5.

232 **"Building in how we"**: Sutton, "Bitter Lesson."

233 "to say that this": Kelion, "DeepMind AI Achieves Grandmaster Status at Starcraft 2."

236 a field of "infinite" play": Derrida, "Structure, Sign, and Play in the Discourse of the Human Sciences," 280.

237 "You are irrelevant and doomed": Gupta, "This AI Chatbot 'Sidney' Is Misbehaving."

239 The biologist Cyrus Levinthal: Levinthal, "How to Fold Graciously."

240 Baker collaborated with game: Cooper et al., "Predicting Protein Structures with a Multiplayer Online Game," 756–60.

242 "When a measure becomes": Strathern, "'Improving Ratings,'" 308.

242 in Scott's terminology: Scott, *Seeing Like a State*, 2.

243 this rise in "deaths of despair": Case and Deaton, *Deaths of Despair*, ix.

245 "a calf thinks God": Rumi, *Essential Rumi*, 223.

245 *ProPublica* reporters analyzed: Angwin et al., "Machine Bias."

246 her thesis on biased computer-vision algorithms: Buolamwini, "Gender Shades," 13–18.

246 As of this writing: Grant and Hill, "Google's Photo App Still Can't Find Gorillas."

246 In 2021, researchers: Bender et al., "On the Dangers of Stochastic Parrots."

Chapter 12: *SimCity*

249 "Human beings, viewed as": Simon, *Sciences of the Artificial*, 53.

249 In *Book of the Customs*: de Cessolis, *Liber de moribus hominum et officiis nobilium ac popularium super ludo scaccorum*.

250 "ought to flee avarice" de Cessolis, *Caxton's Game and Playe of the Chesse, 1474*, 108.

250 "wishes to be greater": Mazour-Matusevich and Korros, *Saluting Aron Gurevich*, 291.

252 From 1970 to 1990: Alfeld, "Urban Dynamics—The First Fifty Years," 207.

252 "the human mind is": Forrester, "Counterintuitive Behavior of Social Systems," 53.

252 "libertarians rather than": Patterson, "Limits of 'Urban Dynamics.'"

255 "All you have to do is adopt": Schlesinger, "SimBill, SimHillary: The High Price of Medi-Carelessness," 118.

255 "*SimHealth* contains so much": Starr, "Seductions of Sim."

255 "just the way the game": Starr.

255 abdicate authority to simulations: Turkle, "Seeing Through Computers."

256 "acted as a site of": Graeber and Wengrow, *Dawn of Everything*, 501.

Chapter 13: Moral Geometry: Playing Utopia

257 "Utopia is the process": Robinson, *Pacific Edge*, 95.

257 "All games have morals": Rushdie, *Midnight's Children*, 160.

258 argues that children's games: Plato, *Laws*.

259 foundation of democracy and of healthy moral: Piaget, *Moral Judgment of the Child*, 366.

260 "no knowledge of the face": Hobbes, *Leviathan* (1904), 83–84.

262 In 1977, game theorist: Ullmann-Margalit, *Emergence of Norms*.

262 **Several scholars have suggested:** Johnson, "God's Punishment and Public Goods"; and Norenzayan, *Big Gods.*
264 **Harsanyi argued that maximin:** Harsanyi, "Cardinal Welfare, Individualistic Ethics, and Interpersonal Comparisons of Utility," 309–21; and Harsanyi, "Can the Maximin Principle Serve as a Basis for Morality?," 594–606.
265 **"a society in which":** Weinstein, "Bringing Logic to Bear on Liberal Dogma."
267 **Skyrms ran simulations:** Skyrms, *Evolution of the Social Contract,* 10–11.
267 **Binmore likens this to:** Binmore, *Natural Justice,* 54.
267 **"he who understands the baboon":** Darwin, *Charles Darwin's Notebooks, 1836–1844,* 539.
267 **"we are indeed naked":** Binmore, *Natural Justice,* 11.
268 **"kind of moral geometry":** Rawls, *Theory of Justice,* 105.
268 **Even other mammals and birds:** Brosnan, "Justice- and Fairness-Related Behaviors in Nonhuman Primates," 10416–23.
268 **"Traditionalists are virulently hostile":** Binmore, *Natural Justice,* 14.
268 **People in many foraging-based:** Binmore, "Justice as a Natural Phenomenon," 9; and Binmore, *Natural Justice,* 40, 130.
269 **In his 1978 book:** Suits, *Grasshopper: Games, Life and Utopia.*
269 **can't miraculously make *any* task pleasant:** Hon, *You've Been Played.*
270 **extols the virtues of games:** McGonigal, *Reality Is Broken.*
270 **Problematic gaming is highest:** André et al., "Gaming Addiction, Problematic Gaming and Engaged Gaming," 100324.
271 **with "monster power":** Green, *Devil's Bargain,* 145.
273 **"art of agency":** Nguyen, *Games,* 17.
274 **The sense of agency:** Lysaker and Leonhardt, "Agency," 165–66.
275 **Researchers at the University:** Kral et al., "Neural Correlates of Video Game Empathy Training in Adolescents," 13.
277 **"I am life which":** Schweitzer, *Out of My Life and Thought,* 156.
277 **"tremendous discovery that ethics":** Schweitzer, *Indian Thought and Its Development,* 83.

Chapter 14: Mechanism Design: Building Games Where Everyone Wins

278 **"Ye live not for yourselves":** Melvill, *Golden Lectures,* 454.
278 **"the waste of all":** Engel, "Virginia Doctor Plans Company to Arrange Sale of Human Kidneys."
279 **"from US citizens and Third World":** Engel.
280 **Board game designer Reiner Knizia:** Chalker, "Reiner Knizia."
280 **created a "kidney clearinghouse":** Roth, Sönmez, and Ünver, "Pairwise Kidney Exchange," 151–88; and Roth, Sönmez, and Ünver, "Kidney Exchange Clearinghouse in New England," 376–80.
281 **It's a hopelessly tangled problem:** von Mises, *Economic Calculation in the Socialist Commonwealth,* 1–46.
284 **In the great chessboard:** Smith, *Theory of Moral Sentiments,* 207.
284 **"Then there is the true":** Holmlund, *Nobel Lectures In Economic Sciences,* 77.
285 **"calls of distress from":** Sidak, *Foreign Investment in American Telecommunications,* 21.
285 **In 1951, the legal scholar:** Herzel, "My 1951 Color Television Article," 523–28.

286 **"To the best of my understanding"**: Rubinstein, *Economic Fables*, 125.

287 **prove collusion definitively:** Cramton and Schwartz, "Collusive Bidding in the FCC Spectrum Auctions," 14.

287 **"mass privatization of a public"**: Weyl, "How Market Design Economists Helped Engineer a Mass Privatization of Public Resources."

289 **"pitcher, batter, and umpire"**: *The State of Texas, et al. v. Google LLC*, 2.

289 **"on both the buy side"**: Horwitz and Hagey, "Google's Secret 'Project Bernanke' Revealed in Texas Antitrust Case."

296 **"I cried myself to sleep"**: Lou, "Facebook Cryptocurrency."

297 **There is a "dark forest"**: Robinson and Konstantopoulos, "Ethereum Is a Dark Forest."

298 **"How did we choose"**: Rubinstein, *Economic Fables*, 217–18.

298 **how humans innovate:** Huizinga, *Homo Ludens*.

299 **"At its deepest"**: Nielsen, "Maps of Matter."

Epilogue

301 **"Man plays only when"**: Schiller, *On the Aesthetic Education of Man*, 80.

301 **"the territory whose shreds"**: Baudrillard, *Jean Baudrillard*, 166.

Bibliography

Aguiar, Mark, Mark Bils, Kerwin Kofi Charles, and Erik Hurst. "Leisure Luxuries and the Labor Supply of Young Men." Working Paper 23552, National Bureau of Economic Research, June 2017. https://doi.org/10.3386/w23552.

Alan of Lille. *Anticlaudianus*. Vol. 3, edited by R. Boussat. Paris: Vrin, 1955.

Alfeld, Louis Edward. "Urban Dynamics—The First Fifty Years." *System Dynamics Review* 11, no. 3 (Fall 1995): 199–217.

Amadae, S. M. *Prisoners of Reason: Game Theory and Neoliberal Political Economy.* 1st ed. Cambridge, UK: Cambridge University Press, 2015. https://doi.org/10.1017/CBO9781107565258.

An, Younggil, and David Ormerod. *Relentless: Lee Sedol vs. Gu Li*. Wilmington, DE: Go Game Guru, 2016.

Anderson, Sam. "Just One More Game . . ." *New York Times Magazine*, April 4, 2012. https://www.nytimes.com/2012/04/08/magazine/angry-birds-farmville-and-other-hyperaddictive-stupid-games.html.

André, Frida, Niroshani Broman, Anders Håkansson, and Emma Claesdotter-Knutsson. "Gaming Addiction, Problematic Gaming and Engaged Gaming—Prevalence and Associated Characteristics." *Addictive Behaviors Reports* 12 (December 2020): 100324. https://doi.org/10.1016/j.abrep.2020.100324.

Andresen, S. L. "Donald Michie: Secrets of Colossus Revealed." *IEEE Intelligent Systems* 16, no. 6 (November 2001): 82–83. https://doi.org/10.1109/5254.972103.

Angwin, Julia, Jeff Larson, Surya Mattu, and Lauren Kirchner. "Machine Bias." *ProPublica*, May 23, 2016. https://www.propublica.org/article/machine-bias-risk-assessments-in-criminal-sentencing.

Anselme, Patrick, and Mike J. F. Robinson. "What Motivates Gambling Behavior? Insight into Dopamine's Role." *Frontiers in Behavioral Neuroscience* 7 (December 2013): 182. https://doi.org/10.3389/fnbeh.2013.00182.

Arbuthnot, John. "An Argument for Divine Providence, Taken from the Constant Regularity Observ'd in the Births of Both Sexes." *Philosophical Transactions of the Royal Society of London* 27, no. 328 (December 1710): 186–90. https://doi.org/10.1098/rstl.1710.0011.

Archibald, Kathleen, ed. *Strategic Interaction and Conflict: Original Papers and Discussion.* Berkeley: Institute of International Studies, University of California, 1966.

Arendt, Hannah. *Crises of the Republic.* New York: Houghton Mifflin Harcourt, 1972.

Aristotle. *Aristotle's Treatise on Poetry, Translated: With Notes on the Translation, and on the Original; and Two Dissertations, on Poetical, and Musical, Imitation.* Vol. 1, translated by Thomas Twining. 2nd ed. London: Luke Hansard & Sons, 1812.

Asimov, Isaac. *The Human Brain, Its Capacities and Functions.* New York: Houghton Mifflin, 1964.

Axelrod, Robert. *The Complexity of Cooperation: Agent-Based Models of Competition and Collaboration.* Princeton, NJ: Princeton University Press, 1997.

Badham, John, dir. *WarGames.* Los Angeles: United Artists and Sherwood Productions, 1983.

Baker, William, and William M. Clarke, eds. *The Letters of Wilkie Collins.* Vol. 2. London: Palgrave Macmillan, 1999.

Barnhart, Russell T. "Gambling in Revolutionary Paris—The Palais Royal: 1789–1838." *Journal of Gambling Studies* 8, no. 2 (June 1992): 151–66. https://doi.org/10.1007/BF01014633.

Barrett, Louise, and Peter Henzi. "The Social Nature of Primate Cognition." *Proceedings of the Royal Society B: Biological Sciences* 272, no. 1575 (September 2005): 1865–75. https://doi.org/10.1098/rspb.2005.3200.

Barrett, Paul H., Peter J. Gautrey, Sandra Herbert, David Kohn, and Sydney Smith, eds. *Charles Darwin's Notebooks, 1836–1844: Geology, Transmutation of Species, Metaphysical Enquiries.* Cambridge, UK: Cambridge University Press, 2008.

Barron, Andrew B., Eirik Søvik, and Jennifer L. Cornish. "The Roles of Dopamine and Related Compounds in Reward-Seeking Behavior Across Animal Phyla." *Frontiers in Behavioral Neuroscience* 4 (October 2010): 163. https://www.frontiersin.org/articles/10.3389/fnbeh.2010.00163.

Baudrillard, Jean. *Jean Baudrillard: Selected Writings.* Edited by Mark Poster. Stanford, CA: Stanford University Press, 1988.

Bellemare, Marc G., Yavar Naddaf, Joel Veness, and Michael Bowling. "The Arcade Learning Environment: An Evaluation Platform for General Agents." *Journal of Artificial Intelligence Research* 47 (June 2013): 253–79. https://doi.org/10.1613/jair.3912.

Bender, Emily M., Timnit Gebru, Angelina McMillan-Major, and Shmargaret Shmitchell. "On the Dangers of Stochastic Parrots: Can Language Models Be Too Big?" In *FAccT '21: Proceedings of the 2021 ACM Conference on Fairness, Accountability, and Transparency,* 610–23. New York: Association for Computing Machinery, 2021. https://doi.org/10.1145/3442188.3445922.

Berliner, Hans J. "Computer Backgammon." *Scientific American* 242, no. 6 (June 1980): 64–72. https://bkgm.com/articles/Berliner/ComputerBackgammon.

Berner, Christopher, Greg Brockman, Brooke Chan, Vicki Cheung, Przemysław Dębiak, Christy Dennison, David Farhi, et al. "Dota 2 with Large-Scale Deep Reinforcement Learning." Preprint, submitted December 13, 2019. https://doi.org/10.48550/arXiv.1912.06680.

Bhargava, Balram, and Rajni Kant. "Health File of Mahatma Gandhi: His Experiments with Dietetics and Nature Cure." *Indian Journal of Medical Research* 149, no. S1 (January 2019): S5–23. https://doi.org/10.4103/0971-5916.251654.

Bhide, Ajit V. "Compulsive Gambling in Ancient Indian Texts." *Indian Journal of Psychiatry* 49, no. 4 (2007): 294–95. https://doi.org/10.4103/0019-5545.37674.

Binmore, Ken. *Game Theory and the Social Contract: Playing Fair.* Cambridge, MA: MIT Press, 1994.

————. "Justice as a Natural Phenomenon." *Analyse & Kritik* 28, no. 1 (2006): 1–12.

————. *Natural Justice.* Oxford: Oxford University Press, 2005.

Birch, Leann Lipps, and Jane Billman. "Preschool Children's Food Sharing with Friends and Acquaintances." *Child Development* 57, no. 2 (April 1986): 387–95. https://doi.org/10.2307/1130594.

Blair, Clay. "Passing of a Great Mind." *Life,* February 25, 1957.

Boden, Margaret. "Obituary: Donald Michie (1923–2007)." *Nature* 448, no. 7155 (August 2007): 765–66. https://doi.org/10.1038/448765a.

Borel, Émile. "On Games That Involve Chance and the Skill of the Players." *Econometrica* 21, no. 1 (January 1953): 101–15. https://doi.org/10.2307/1906947.

Borges, Jorge Luis. *Collected Fictions.* Translated by Andrew Hurley. New York: Penguin Books, 1999.

Bott, Raoul. *Mathematics Related to Physics.* New York: Springer Science & Business Media, 1994.

Brisch, Ralf, Arthur Saniotis, Rainer Wolf, Hendrik Bielau, Hans-Gert Bernstein, Johann Steiner, Bernhard Bogerts, et al. "The Role of Dopamine in Schizophrenia from a Neurobiological and Evolutionary Perspective: Old Fashioned, but Still in Vogue." *Frontiers in Psychiatry* 5 (May 2014): 47. https://www.frontiersin.org/articles/10.3389/fpsyt.2014.00047.

British Go Association. "History of Go-Playing Programs." Updated January 2, 2018. https://www.britgo.org/computergo/history.

Brodie, Bernard. "Why Were We So (Strategically) Wrong?" *Foreign Policy,* no. 5 (Winter 1971–72): 151–61. https://doi.org/10.2307/1147725.

Brosnan, Sarah F. "Justice- and Fairness-Related Behaviors in Nonhuman Primates." *Proceedings of the National Academy of Sciences* 110, no. S2 (June 2013): 10416–23. https://doi.org/10.1073/pnas.1301194110.

Brownell, Celia A., Margarita Svetlova, and Sara Nichols. "To Share or Not to Share: When Do Toddlers Respond to Another's Needs?" *Infancy: The Official Journal of the International Congress of Infant Studies* 14, no. 1 (2009): 117–30. https://doi.org/10.1080/15250000802569868.

Brügmann, Bernd. "Monte Carlo Go." Max Planck Institute of Physics, Munich, Germany, October 9, 1993. http://www.ideanest.com/vegos/MonteCarloGo.pdf.

Buolamwini, Joy Adowaa. "Gender Shades: Intersectional Phenotypic and Demographic Evaluation of Face Datasets and Gender Classifiers." Master's thesis, MIT, 2017. https://www.media.mit.edu/publications/full-gender-shades-thesis-17.

Burke, Christopher J., Alexander Soutschek, Susanna Weber, Anjali Raja Beharelle, Ernst Fehr, Helene Haker, and Philippe N. Tobler. "Dopamine Receptor-Specific Contributions to the Computation of Value." *Neuropsychopharmacology* 43, no. 6 (May 2018): 1415–24. https://doi.org/10.1038/npp.2017.302.

Bushnell, Nolan. "E29: Nolan Bushnell, Founder Chairman at Snap Institute and Founder of Atari—Interview." By Dave Kruse. *Flyover Labs* (podcast), June 7, 2016. SoundCloud audio, 21:44. https://www.flyoverlabs.io/podcasts/nolan-bushnell-founder-chairman-at-snap-institute-and-founder-of-atari-interview.

Caffrey, Matthew, Jr. "Toward a History-Based Doctrine for Wargaming." *Aerospace Power Journal* 14, no. 3 (Fall 2000): 33–56. https://apps.dtic.mil/sti/pdfs/ADA 521381.pdf.

Calado, Filipa, and Mark D. Griffiths. "Problem Gambling Worldwide: An Update and Systematic Review of Empirical Research (2000–2015)." *Journal of Behavioral Addictions* 5, no. 4 (October 2016): 592–613. https://doi.org/10.1556/2006.5 .2016.073.

Cardano, Girolamo. *The Book of My Life*. Translated by Jean Stoner. New York: E. P. Dutton & Co., 1930.

Carlsson, Arvid, Margit Lindqvist, and Tor Magnusson. "3,4-Dihydroxyphenylalanine and 5-Hydroxytryptophan as Reserpine Antagonists." *Nature* 180, no. 4596 (1957): 1200. https://doi.org/10.1038/1801200a0.

Case, Anne and Angus Deaton. *Deaths of Despair and the Future of Capitalism*. Princeton, NJ: Princeton University Press, 2021.

Chalker, Dave. "Reiner Knizia: 'Creation of a Successful Game.'" *Critical-Hits* (blog), July 3, 2008. https://critical-hits.com/blog/2008/07/03/reiner-knizia-creation-of -a-successful-game.

Cooper, Seth, Firas Khatib, Adrien Treuille, Janos Barbero, Jeehyung Lee, Michael Beenen, Andrew Leaver-Fay, David Baker, Zoran Popović, and Foldit Players. "Predicting Protein Structures with a Multiplayer Online Game." *Nature* 466, no. 7307 (August 2010): 756–60. https://doi.org/10.1038/nature09304.

Costikyan, Greg. *Uncertainty in Games*. Playful Thinking Series. Cambridge, MA: MIT Press, 2013.

Cottrell, G. A. "Occurrence of Dopamine and Noradrenaline in the Nervous Tissue of Some Invertebrate Species." *British Journal of Pharmacology and Chemotherapy* 29, no. 1 (January 1967): 63–69. https://doi.org/10.1111/j.1476-5381.1967.tb01 939.x.

Cournot, Augustin. *Researches into the Mathematical Principles of the Theory of Wealth*. Translated by Nathaniel T. Bacon. New York: Macmillan, 1897.

Craig, Hardin. "Hamlet's Book." *Huntington Library Bulletin*, no. 6 (November 1934): 17–37. https://www.jstor.org/stable/3818164.

Cramton, Peter, and Jesse A Schwartz. "Collusive Bidding in the FCC Spectrum Auctions." *B.E. Journal of Economic Analysis & Policy* 1, no. 1 (December 2001). https:// doi.org/10.2202/1538-0645.1078.

Culin, Stewart. "American Indian Games (1902)." *American Anthropologist* 5, no. 1 (1903): 58–64. https://www.jstor.org/stable/659360.

Cummings, E. E. *Complete Poems, 1904–1962*. Edited by George J. Firmage. Rev. ed. New York: Liveright, 1991.

D'Agostino, Susan. "DeepMind's David Silver on Games, Beauty, and AI's Potential to Avert Human-Made Disasters." *Bulletin of the Atomic Scientists* (blog), January 10, 2022. https://thebulletin.org/2022/01/deepminds-david-silver-on-games-beauty -and-ais-potential-to-avert-human-made-disasters.

Darwin, Charles. *The Autobiography of Charles Darwin, 1809–1882*. Edited by Nora Barlow. New York: W. W. Norton & Company, 1969.

———. *The Life and Letters of Charles Darwin: Including an Autobiographical Chapter*. Edited by Francis Darwin. 2 vols. New York: D. Appleton and Company, 1896.

———. *The Origin of Species by Means of Natural Selection, or the Preservation of*

Favored Races in the Struggle for Life, and the Descent of Man and Selection in Relation to Sex. New York: Modern Library, 1872.

Daubin, Vincent, and Gergely J. Szöllösi. "Horizontal Gene Transfer and the History of Life." *Cold Spring Harbor Perspectives in Biology* 8, no. 4 (April 2016): a018036. https://doi.org/10.1101/cshperspect.a018036.

David, Florence Nightingale. *Games, Gods and Gambling: The Origins and History of Probability and Statistical Ideas from the Earliest Times to the Newtonian Era*. New York: Hafner Publishing Company, 1962.

Davis, Nuel Pharr. *Lawrence and Oppenheimer*. New York: Simon & Schuster, 1968.

Dawkins, Richard. *The Selfish Gene*. 30th anniversary ed. Oxford: Oxford University Press, 2006.

de Bloch, Jean. *The Future of War in Its Technical, Economic, and Political Relations*. Translated by R. C. Long. Boston: World Peace Foundation, 1914.

de Cervantes, Miguel. *Don Quixote*. Translated by Edith Grossman. New York: Harper-Collins, 2009.

de Cessolis, Jacobus. *Caxton's Game and Playe of the Chesse, 1474*. Translated by William Caxton. London: Elliot Stock. 1883.

———. *Liber de moribus hominum et officiis nobilium ac popularium super ludo scaccorum*, c. 1390–1410. Corpus Christi College, Cambridge University, MS 177.

De Herdt, Tom. "Cooperation and Fairness: The Flood–Dresher Experiment Revisited." *Review of Social Economy* 61, no. 2 (June 2003): 183–210. https://doi.org /10.1080/0034676032000098219.

DeBoer, Warren R. "Of Dice and Women: Gambling and Exchange in Native North America." *Journal of Archaeological Method and Theory* 8, no. 3 (September 2001): 215–68. https://www.jstor.org/stable/20177442.

Degenhardt, Louisa, Fiona Charlson, Bradley Mathers, Wayne D. Hall, Abraham D. Flaxman, Nicole Johns, and Theo Vos. "The Global Epidemiology and Burden of Opioid Dependence: Results from the Global Burden of Disease 2010 Study." *Addiction* 109, no. 8 (2014): 1320–33. https://doi.org/10.1111/add.12551.

Delgado, José M. R. *Physical Control of the Mind: Toward a Psychocivilized Society*. New York: Harper & Row, 1969.

Derrida, Jacques. "Structure, Sign, and Play in the Discourse of the Human Sciences." In *Writing and Difference*, translated by Alan Bass, 278–93. London: Routledge, 1978.

Dijkstra, Edsger W. *Selected Writings on Computing: A Personal Perspective*. New York: Springer-Verlag, 1982.

Dorsey, George Amos. *Why We Behave Like Human Beings*. New York: Harper & Brothers, 1925.

Dotson, Brandon, Constance A. Cook, and Zhao Lu. *Dice and Gods on the Silk Road: Chinese Buddhist Dice Divination in Transcultural Context*. Prognostication in History 7. Leiden and Boston: Brill, 2021.

Dudley, Brier. "The Business of Gaming: It's Bigger than You Might Think, and It's Growing." *University of Washington Magazine*. Accessed April 19, 2023. https:// magazine.washington.edu/feature/the-business-of-gaming-its-bigger-than-you -might-think-and-its-growing.

Duffin, Richard J. "Nash, John Forbes, Jr., 1950," 1948. Nash, John Forbes, Jr., 1950. Graduate Alumni Records, 1930–1959. Princeton University Archives, Depart-

ment of Special Collections, Princeton University Library. https://findingaids
.princeton.edu/catalog/AC105-03_c3166.

Dunbar, R. I. M. "Neocortex Size as a Constraint on Group Size in Primates." *Journal of Human Evolution* 22, no. 6 (June 1992): 469–93. https://doi.org/10.1016/0047
-2484(92)90081-J.

———. "The Social Brain Hypothesis and Its Implications for Social Evolution." *Annals of Human Biology* 36, no. 5 (2009): 562–72. https://doi.org/10.1080
/03014460902960289.

Dusaulx, Jean. *De la passion du jeu, depuis les temps anciens jusqu'a nos jours.* Paris: Imprimerie de Monsieur, 1779.

Dyson, George. *Turing's Cathedral: The Origins of the Digital Universe.* New York: Pantheon Books, 2012.

Edwards, Benj. "Robots, Pizza, and Sensory Overload: The Chuck E. Cheese Origin Story." *Fast Company*, May 31, 2017. https://www.fastcompany.com/40425172
/robots-pizza-and-magic-the-chuck-e-cheese-origin-story.

Eisenhower, Dwight D. "President Dwight D. Eisenhower's Farewell Address." Transcript of speech delivered at the White House, Washington, DC, January 17, 1961. https://www.archives.gov/milestone-documents/president-dwight-d-eisenhowers
-farewell-address.

Engel, Margaret. "Virginia Doctor Plans Company to Arrange Sale of Human Kidneys." *Washington Post*, September 19, 1983. https://www.washingtonpost.com
/archive/politics/1983/09/19/va-doctor-plans-company-to-arrange-sale-of
-human-kidneys/afdfac69-62ed-4066-b296-fcf892eab758.

Erwin, Douglas. "A Call to the Custodians of Deep Time." *Nature* 462, no. 7271 (November 2009): 282–83. https://doi.org/10.1038/462282a.

Fennell, Lee Anne. "Ostrom's Law: Property Rights in the Commons." *International Journal of the Commons* 5, no. 1 (2011): 9–27. https://doi.org/10.18352/ijc.252.

Feynman, Richard P., Robert B. Leighton, and Matthew Sands. *The Feynman Lectures on Physics.* New millennium ed. New York: Basic Books, 2011.

Fisher, R. A. *Collected Papers of R. A. Fisher.* Edited by J. H. Bennett. Vol. 4. Adelaide, AU: University of Adelaide Press, 1974.

FORA.tv, "Dopamine Jackpot! Sapolsky on the Science of Pleasure." Streamed on March 2, 2011. YouTube video, 4:59. https://www.youtube.com/watch?v=axry
wDP9Ii0.

Forrester, Jay W. "Counterintuitive Behavior of Social Systems." *Technology Review* 73, no. 3 (January 1971): 52–68.

Foster, Meadhbh I., and Mark T. Keane. "The Role of Surprise in Learning: Different Surprising Outcomes Affect Memorability Differentially." *Topics in Cognitive Science* 11, no. 1 (January 2019): 75–87. https://doi.org/10.1111/tops.12392.

France, Clemens J. "The Gambling Impulse." *American Journal of Psychology* 13, no. 3 (July 1902): 364–407. https://doi.org/10.2307/1412559.

Frank, Steven A. "The Price Equation Program: Simple Invariances Unify Population Dynamics, Thermodynamics, Probability, Information and Inference." *Entropy* 20, no. 12 (December 2018): e20120978. https://doi.org/10.3390/e20120978.

Fréchet, Maurice. "Commentary on the Three Notes of Emile Borel." *Econometrica* 21, no. 1 (January 1953): 118–24. https://doi.org/10.2307/1906949.

Galton, Francis. *Memories of My Life*. London: Methuen & Company, 1908.

Gardner, Martin. "Mathematical Games." *Scientific American* 242, no. 1 (January 1980): 22–33B. https://www.jstor.org/stable/e24966221.

George, Sanju. "From the Gambler Within: Dostoyevsky's *The Gambler.*" *Advances in Psychiatric Treatment* 18, no. 3 (May 2012): 226–31. https://doi.org/10.1192/apt.bp.111.008995.

Gintis, Herbert. "Five Principles for the Unification of the Behavioral Sciences." Working paper, Santa Fe Institute, May 13, 2008. https://www.researchgate.net/profile/Herbert-Gintis/publication/265566720_Five_Principles_for_the_Unification_of_the_Behavioral_Sciences/links/5513edfb0cf283ee083497e3/Five-Principles-for-the-Unification-of-the-Behavioral-Sciences.pdf.

———. *The Bounds of Reason: Game Theory and the Unification of the Behavioral Sciences*. Princeton, NJ: Princeton University Press, 2009.

Glimm, James, John Impagliazzo, and Isadore Singer, eds. *The Legacy of John von Neumann*. Proceedings of Symposia in Pure Mathematics, vol. 50. Providence, RI: American Mathematical Society, 2006.

Goodwin, Greenville. *The Social Organization of the Western Apache*. Tucson, AZ: University of Arizona Press, 1969. https://open.uapress.arizona.edu/projects/the-social-organization-of-the-western-apache.

Google DeepMind. *AlphaGo—The Movie*. Streamed on March 13, 2020. YouTube video, 1:30:27. https://www.youtube.com/watch?v=WXuK6gekU1Y.

Gopnik, Alison. *The Philosophical Baby: What Children's Minds Tell Us About Truth, Love, and the Meaning of Life*. New York: Farrar, Straus and Giroux, 2009.

Graeber, David, and David Wengrow. *The Dawn of Everything: A New History of Humanity*. New York: Farrar, Straus and Giroux, 2021.

Grant, Nico, and Kashmir Hill. "Google's Photo App Still Can't Find Gorillas. And Neither Can Apple's." *New York Times*, May 22, 2023. https://www.nytimes.com/2023/05/22/technology/ai-photo-labels-google-apple.html.

Green, Joshua. *Devil's Bargain: Steve Bannon, Donald Trump, and the Storming of the Presidency*. New York: Penguin Press, 2017.

Greggor, Alison L., Bryce Masuda, Jacqueline M. Gaudioso-Levita, Jay T. Nelson, Thomas H. White, Debra M. Shier, Susan M. Farabaugh, and Ronald R. Swaisgood. "Pre-release Training, Predator Interactions and Evidence for Persistence of Anti-predator Behavior in Reintroduced 'Alalā, Hawaiian Crow." *Global Ecology and Conservation* 28 (August 2021): e01658. https://doi.org/10.1016/j.gecco.2021.e01658.

Gregory, Anne. "Chunkey, Cahokia, and Indigenous Conflict Resolution." Master's thesis, University of Oregon, June 2020. https://scholarsbank.uoregon.edu/xmlui/handle/1794/25664.

Groos, Karl. *The Play of Animals*. Translated by Elizabeth L. Baldwin. New York: D. Appleton and Company, 1898. http://archive.org/details/playofanimals00groouoft.

Gupta, Deepa. "This AI Chatbot 'Sidney' Is Misbehaving." Microsoft Community Answers, November 23, 2022. https://answers.microsoft.com/en-us/bing/forum/all/this-ai-chatbot-sidney-is-misbehaving/e3d6a29f-06c9-441c-bc7d-51a68e856761.

Haldeman, Harry R., and Joseph DiMona. *The Ends of Power*. New York: Times Books, 1978.

Halmos, P. R. "The Legend of John von Neumann." *American Mathematical Monthly* 80, no. 4 (April 1973): 382–94. https://doi.org/10.2307/2319080.

Hamilton, Alexander, James Madison, and John Jay. *The Federalist Papers*. New York: Signet, 2003.

Hamilton, W. D. "Extraordinary Sex Ratios." *Science* 156, no. 3774 (April 1967): 477–88. https://doi.org/10.1126/science.156.3774.477.

———."The Genetical Evolution of Social Behaviour. I." *Journal of Theoretical Biology* 7, no. 1 (July 1964): 1–16. https://doi.org/10.1016/0022-5193(64)90038-4.

Hannak, J. *Emanuel Lasker: The Life of a Chess Master*. Mineola, NY: Courier Dover Publications, 1991.

Hardin, Garrett. "Living in a World of Limits—An Interview with Noted Biologist Garrett Hardin." By Craig Straub. Garrett Hardin Society, Fall 1997. https://www.garretthardinsociety.org/gh/gh_straub_interview.html.

———. "The Tragedy of the Commons." *Science* 162, no. 3859 (December 1968): 1243–48. https://doi.org/10.1126/science.162.3859.1243.

Harman, Oren. *The Price of Altruism: George Price and the Search for the Origins of Kindness*. New York: W. W. Norton & Company, 2011.

Harsanyi, John C. "Can the Maximin Principle Serve as a Basis for Morality? A Critique of John Rawls's Theory." *American Political Science Review* 69, no. 2 (1975): 594–606. https://doi.org/10.2307/1959090.

———. "Cardinal Welfare, Individualistic Ethics, and Interpersonal Comparisons of Utility." *Journal of Political Economy* 63, no. 4 (August 1955): 309–21. https://www.jstor.org/stable/1827128.

Hartston, William. "Can God Beat the Computer?: Today the World's Greatest Draughts Player Faces a Challenge from an Electronic Brain, Says William Hartston." *Independent*, August 16, 1992. https://www.independent.co.uk/life-style/can-god-beat-the-computer-today-the-world-s-greatest-draughts-player-faces-a-challenge-from-an-electronic-brain-says-william-hartston-1540861.html.

———. "Particularly Susceptible to Draughts." *Independent*, July 18, 1997. https://www.independent.co.uk/life-style/particularly-susceptible-to-draughts-1251328.html.

Haydon, Benjamin Robert. *The Life, Letters and Table Talk of Benjamin Robert Haydon*. Edited by Richard Henry Stoddard. New York: Scribner, Armstrong and Company, 1876.

Heiden, Petra, Andreas Heinz, and Nina Romanczuk-Seiferth. "Pathological Gambling in Parkinson's Disease: What Are the Risk Factors and What Is the Role of Impulsivity?" *European Journal of Neuroscience* 45, no. 1 (January 2017): 67–72. https://doi.org/10.1111/ejn.13396.

Herodotus. *Herodotus: A New and Literal Version from the Text of Baehr*. Translated by Henry Cary. London: H. G. Bohn, 1848.

Herzel, Leo. "My 1951 Color Television Article." *Journal of Law & Economics* 41, no. S2 (October 1998): 523–28. https://doi.org/10.1086/467401.

Hesse, Hermann. *The Glass Bead Game: (Magister Ludi)*. New York: Picador, 2002.

Hilbert, David. "The Foundation of Mathematics." In *The Emergence of Logical Empiricism: From 1900 to the Vienna Circle*, edited by Sahotra Sarkar, 228–43. New York: Garland, 1996.

———. "The New Grounding of Mathematics: First Report." In *From Brouwer to Hilbert: The Debate on the Foundations of Mathematics in the 1920s*, edited by Paolo Mancosu, 198–214. Oxford: Oxford University Press, 1998.

Hitch, Charles. "Uncertainties in Operations Research." *Operations Research* 8, no. 4 (1960): 437–45.

Hobbes, Thomas. *Leviathan*. North Chelmsford, MA: Courier Corporation, 2012.

———. *Leviathan, or the Matter, Forme & Power of a Commonwealth, Ecclesiasticall and Civill*. Edited by A. R. Waller. Cambridge, UK: Cambridge University Press, 1904.

Hoffman, Leslie A., and Joel A. Vilensky. "Encephalitis Lethargica: 100 Years after the Epidemic." *Brain* 140, no. 8 (August 2017): 2246–51. https://doi.org/10.1093/brain/awx177.

Holmlund, Bertil. *Nobel Lectures in Economic Sciences (2006–2010)*. Singapore: World Scientific, 2014.

Hon, Adrian. *You've Been Played: How Corporations, Governments, and Schools Use Games to Control Us All*. New York: Basic Books, 2022.

Hopp, Toby, Scott Parrott, and Yuan Wang. "Use of Military-Themed First-Person Shooter Games and Militarism: An Investigation of Two Potential Facilitating Mechanisms." *Computers in Human Behavior* 78 (January 2018): 192–99. https://doi.org/10.1016/j.chb.2017.09.035.

Hornykiewicz, Oleh. "Dopamine Miracle: From Brain Homogenate to Dopamine Replacement." *Movement Disorders* 17, no. 3 (2002): 501–8. https://doi.org/10.1002/mds.10115.

Horwitz, Jeff, and Keach Hagey. "Google's Secret 'Project Bernanke' Revealed in Texas Antitrust Case." *Wall Street Journal*, April 10, 2021. https://www.wsj.com/articles/googles-secret-project-bernanke-revealed-in-texas-antitrust-case-11618097760.Hsu, Feng-hsiung. *Behind Deep Blue: Building the Computer That Defeated the World Chess Champion*. Princeton, NJ: Princeton University Press, 2022.

Huizinga, Johan. *Homo Ludens*. New York: Random House, 1938.

Huygens, Christiaan. *Christiani Hugenii libellus de ratiociniis in ludo aleae. Or, the Value of All Chances in Games of Fortune; Cards, Dice, Wagers, Lotteries, &c. Mathematically Demonstrated*. London: T. Woodward, 1714.

Johnson, Daniel. *White King and Red Queen: How the Cold War Was Fought on the Chessboard*. New York: Houghton Mifflin, 2008.

Johnson, Dominic D. P. "God's Punishment and Public Goods : A Test of the Supernatural Punishment Hypothesis in 186 World Cultures." *Human Nature* 16, no. 4 (December 2005): 410–46. https://doi.org/10.1007/s12110-005-1017-0.

Jones, Ernest. "The Problem of Paul Morphy: A Contribution to the Psycho-Analysis of Chess." *International Journal of Psycho-Analysis* 12 (January 1931). http://www.edochess.ca/batgirl/Jones.html.

Jorgensen, Janice, ed. *Durable Goods*. Vol. 3 of *Encyclopedia of Consumer Brands*. Farmington Hills, MI: St. James Press, 1994.

Kahn, Herman. *In Defense of Thinking: The Essential Herman Kahn*. Edited by Paul Dragos Aligica and Kenneth R. Weinstein. Lanham, MD: Lexington Books, 2009.

———. *Thinking about the Unthinkable in the 1980s*. New York: Touchstone Books, 1985.

Karnow, Stanley. *Vietnam: A History*. New York: Penguin Books, 1997.

Kasparov, Garry. "Garry Kasparov on Chess and Politics in Soviet Russia." Interview by Bill Kristol. *Conversations with Bill Kristol*, February 29, 2016. https://conversa tionswithbillkristol.org/transcript/garry-kasparov-transcript.

Kass-Simon, G., and Paola Pierobon. "Cnidarian Chemical Neurotransmission, an Updated Overview." *Comparative Biochemistry and Physiology Part A: Molecular & Integrative Physiology* 146, no. 1 (January 2007): 9–25. https://doi.org/10.1016 /j.cbpa.2006.09.008.

Keeny, Spurgeon M., and Wolfgang K. H. Panofsky. "Mad versus Nuts: Can Doctrine or Weaponry Remedy the Mutual Hostage Relationship of the Superpowers?" *Foreign Affairs* 60, no. 2 (Winter 1981): 287–304. https://doi.org/10.2307/20041081.

Kelion, Leo. "DeepMind AI Achieves Grandmaster Status at Starcraft 2." *BBC News*, October 30, 2019. https://www.bbc.com/news/technology-50212841.

Keynes, Milo, and G. Ainsworth Harrison, eds. *Evolutionary Studies: A Centenary Celebration of the Life of Julian Huxley*. Studies in Biology, Economy and Society. London: Palgrave Macmillan, 1989.

Kindt, Katie S., Kathleen B. Quast, Andrew C. Giles, Subhajyoti De, Dan Hendrey, Ian Nicastro, Catharine H. Rankin, and William R. Schafer. "Dopamine Mediates Context-Dependent Modulation of Sensory Plasticity in C. *elegans.*" *Neuron* 55, no. 4 (August 2007): 662–76. https://doi.org/10.1016/j.neuron.2007.07.023.

Klopf, A. Harry. *Brain Function and Adaptive Systems: A Heterostatic Theory*. Special Report No. 133. Bedford, MA: Air Force Cambridge Research Laboratories, 1972. https://apps.dtic.mil/sti/pdfs/AD0742259.pdf.

Kniffin, Kevin M., and Michelle Scalise Sugiyama. "Toward a Natural History of Team Sports." *Human Nature* 29, no. 3 (September 2018): 211–18. https://doi.org /10.1007/s12110-018-9322-6.

Kobayashi, Kenji, and Ming Hsu. "Common Neural Code for Reward and Information Value." *Proceedings of the National Academy of Sciences* 116, no. 26 (June 2019): 13061–66. https://doi.org/10.1073/pnas.1820145116.

Koster, Raph. *A Theory of Fun for Game Design*. Sebastopol, CA: O'Reilly Media, 2013.

Kral, Tammi R. A., Diane E. Stodola, Rasmus M. Birn, Jeanette A. Mumford, Enrique Solis, Lisa Flook, Elena G. Patsenko, Craig G. Anderson, Constance Steinkuehler, and Richard J. Davidson. "Neural Correlates of Video Game Empathy Training in Adolescents: A Randomized Trial." *npj Science of Learning* 3, no. 1 (2018): 13. https://doi.org/10.1038/s41539-018-0029-6.

Krauthammer, Charles. "Be Afraid." *Washington Examiner*, May 26, 1997. https:// www.washingtonexaminer.com/weekly-standard/be-afraid-9802.

Kuchynka, Sophie L., Jennifer K. Bosson, Joseph A. Vandello, and Curtis Puryear. "Zero-Sum Thinking and the Masculinity Contest: Perceived Intergroup Competition and Workplace Gender Bias." *Journal of Social Issues* 74, no. 3 (September 2018): 529–50. https://doi.org/10.1111/josi.12281.

Kumar, Rahul, and Vikram Yeragani. "Penfield—A Great Explorer of Psyche-Soma-Neuroscience." *Indian Journal of Psychiatry* 53, no. 3 (July 1, 2011): 276–78.

Kuperczko, Diana, Peter Kenyeres, Gergely Darnai, Norbert Kovacs, and Jozsef Janszky. "Sudden Gamer Death: Non-violent Death Cases Linked to Playing Video Games." *BMC Psychiatry* 22, no. 1 (December 2022): 824. https://doi.org/10.1186 /s12888-022-04373-5.

Lanciani, Rodolfo. "Gambling and Cheating in Ancient Rome." *North American Review* 155, no. 428 (July 1892): 97–105. https://www.jstor.org/stable/25102412.

Laplace, Pierre-Simon. *A Philosophical Essay on Probabilities*. Translated by Frederick Wilson Truscott and Frederick Lincoln Emory. New York: John Wiley & Sons, 1902.

Larson, Laurence Marcellus. *Canute the Great 995 (circ)—1035 and the Rise of Danish Imperialism During the Viking Age*. New York: G. P. Putnam's Sons, 1912.

Lasker, Emanuel. *Lasker's Manual of Chess*. Milford, CT: Russell Enterprises, Inc., 2010.

———. *Struggle*. New York: Lasker's Publishing Company, 1907.

Lee, Stephanie M. "A Weird Research-Misconduct Scandal about Dishonesty Just Got Weirder." *Chronicle of Higher Education*, June 16, 2023. https://www.chronicle.com/article/a-weird-research-misconduct-scandal-about-dishonesty-just-got-weirder.

Leeson, Bill. "The Origins of Kriegsspiel." *Kriegsspiel* (blog), accessed October 17, 2023. https://kriegsspielorg.wordpress.com/home/the-origins-of-kriegsspiel.

Leibniz, Gottfried Wilhelm. "A Note on Certain Games, Especially a Certain Chinese Game, and on the Difference between Chess and Latrunculus and on a New Kind of Naval Game." *Miscellanea Berolinensia* 1 (1710): 22–26.

———. *Die philosophischen Schriften von Gottfried Wilhelm Leibniz*. Edited by C. I. Gerhardt. 7 vols. Hildesheim: G. Olms, 1960.

Leibo, Joel Z., Edward Hughes, Marc Lanctot, and Thore Graepel. "Autocurricula and the Emergence of Innovation from Social Interaction: A Manifesto for Multi-agent Intelligence Research." Preprint, submitted March 2, 2019. http://arxiv.org/abs/1903.00742.

Leonard, Robert. *Von Neumann, Morgenstern, and the Creation of Game Theory: From Chess to Social Science, 1900–1960*. Historical Perspectives on Modern Economics. Cambridge, UK: Cambridge University Press, 2010.

Leonard, Robert J. "From Parlor Games to Social Science: Von Neumann, Morgenstern and the Creation of Game Theory, 1928–1944." *Journal of Economic Literature* 33, no. 2 (June 1995): 730–61.

Levinthal, C. "How to Fold Graciously." Lecture presented at the Mössbaun Spectroscopy in Biological Systems Proceedings, Urbana, IL, 1969. https://www.semanticscholar.org/paper/How-to-fold-graciously-Levinthal/1ef89dfb1e3404f4ace99399ce582b2bc982d0bf.

Liberman, Varda, Steven M. Samuels, and Lee Ross. "The Name of the Game: Predictive Power of Reputations versus Situational Labels in Determining Prisoner's Dilemma Game Moves." *Personality and Social Psychology Bulletin* 30, no. 9 (September 2004): 1175–85. https://doi.org/10.1177/0146167204264004.

Lihoreau, Mathieu, Tanya Latty, and Lars Chittka. "An Exploration of the Social Brain Hypothesis in Insects." *Frontiers in Physiology* 3 (November 2012): 442. https://www.frontiersin.org/articles/10.3389/fphys.2012.00442.

Lipkowitz, Elise. "The Physicians' Dilemma in the 18th-Century French Smallpox Debate." *Journal of the American Medical Association* 290, no. 17 (November 2003): 2329–30. https://doi.org/10.1001/jama.290.17.2329.

Lou, Ethan. "Facebook Cryptocurrency: Dawn of the Corporation-Government." *Guardian*, June 19, 2019. https://www.theguardian.com/commentisfree/2019/jun /19/facebook-cryptocurrency-libra.

Lovelace, Augusta Ada. "On Babbage's Analytical Engine: Translator's Notes to M. Menabrea's Memoir." In *Scientific Memoirs, Selected from the Transactions of Foreign Academies of Science and Learned Societies, and from Foreign Journals*. Vol. 3, edited by Richard Taylor, 691–731. London: Richard and John E. Taylor, 1843.

Lysaker, Paul H., and Bethany L. Leonhardt. "Agency: Its Nature and Role in Recovery from Severe Mental Illness." *World Psychiatry* 11, no. 3 (October 2012): 165–66. https://doi.org/10.1002/j.2051-5545.2012.tb00121.x.

MacAskill, William. *What We Owe the Future*. New York: Basic Books, 2022.

Maia, Tiago V., and Michael J. Frank. "An Integrative Perspective on the Role of Dopamine in Schizophrenia." *Biological Psychiatry* 81, no. 1 (January 2017): 52–66. https://doi.org/10.1016/j.biopsych.2016.05.021.

Marsden, Charles A. "Dopamine: The Rewarding Years." *British Journal of Pharmacology* 147, no. S1 (January 2006): S136–44. https://doi.org/10.1038/sj.bjp.0706473.

Marx, Karl. *Capital*. Vol. 1, *A Critique of Political Economy*, edited by Friedrich Engels and translated by Samuel Moore and Edward Aveling. Mineola, NY: Courier Dover Publications, 2019.

Maugham, W. Somerset. *Strictly Personal*. Garden City: Doubleday, Doran and Company, 1941.

Maurer, Petra H., Donatella Rossi, and Rolf Scheuermann, eds. *Glimpses of Tibetan Divination: Past and Present*. Prognostication in History. Leiden: Brill, 2020.

Maynard Smith, John. *On Evolution*. Edinburgh: Edinburgh University Press, 1972.

Mazour-Matusevich, Yelena, and Alexandra S. Korros. *Saluting Aron Gurevich: Essays in History, Literature and Other Related Subjects*. Leiden: Brill, 2010.

McCarthy, J. "Chess as the Drosophila of AI." In *Computers, Chess, and Cognition*, edited by T. Anthony Marsland and Jonathan Schaeffer, 227–37. New York: Springer-Verlag, 1990. https://doi.org/10.1007/978-1-4613-9080-0_14.

McCarthy, John. "What Is Artificial Intelligence?" John McCarthy's Web Page (Stamford). Revised November 12, 2007. http://www-formal.stanford.edu/jmc/whati sai/whatisai.html.

McGhee, Heather. *The Sum of Us: What Racism Costs Everyone and How We Can Prosper Together*. New York: Random House, 2021.

McGonigal, Jane. *Reality Is Broken: Why Games Make Us Better and How They Can Change the World*. New York: Penguin Press, 2011.

Melvill, Henry. *The Golden Lectures: Forty-Five Sermons Delivered at St. Margaret's Church, Lothbury, on Tuesday Mornings, from January 2 to December 18, 1855*. The Preacher in Print. London, 1855.

Menzies, William Cameron, dir. *Things to Come*. London: London Film Productions, 1936.

Metropolis, Nicholas, Arianna W. Rosenbluth, Marshall N. Rosenbluth, Augusta H. Teller, and Edward Teller. "Equation of State Calculations by Fast Computing Machines." *Journal of Chemical Physics* 21, no. 6 (1953): 1087–92. https://doi.org /10.1063/1.1699114.

Metz, Cade. "The Sadness and Beauty of Watching Google's AI Play Go." *Wired*, March 11, 2016. https://www.wired.com/2016/03/sadness-beauty-watching-googles-ai-play-go.

Michie, Donald. "Experiments on the Mechanization of Game-Learning Part I. Characterization of the Model and Its Parameters." *Computer Journal* 6, no. 3 (November 1963): 232–36. https://doi.org/10.1093/comjnl/6.3.232.

Millard, Samuel J., Carrie E. Bearden, Katherine H. Karlsgodt, and Melissa J. Sharpe. "The Prediction-Error Hypothesis of Schizophrenia: New Data Point to Circuit-Specific Changes in Dopamine Activity." *Neuropsychopharmacology* 47, no. 3 (2022): 628–40. https://doi.org/10.1038/s41386-021-01188-y.

Miller, Andrew M., trans., *Greek Lyric: An Anthology in Translation*. Indianapolis: Hackett Publishing, 1996.

Miller, Lee A., and Annemarie Surlykke. "How Some Insects Detect and Avoid Being Eaten by Bats: Tactics and Countertactics of Prey and Predator." *BioScience* 51, no. 7 (July 2001): 570–81. https://doi.org/10.1641/0006-3568(2001)051[0570:HSIDAA]2.0.CO;2.

Mills, Robin, Graham K. Taylor, and Charlotte K. Hemelrijk. "Sexual Size Dimorphism, Prey Morphology and Catch Success in Relation to Flight Mechanics in the Peregrine Falcon: A Simulation Study." *Journal of Avian Biology* 50, no. 3 (March 2019): e01979. https://doi.org/10.1111/jav.01979.

Milne, David. *America's Rasputin: Walt Rostow and the Vietnam War*. New York: Macmillan, 2009.

Mirowski, Philip. "What Were von Neumann and Morgenstern Trying to Accomplish?" In *Toward a History of Game Theory*, edited by E. Roy Weintraub, 111–47. Durham, NC: Duke University Press, 1992.

Moan, Charles E., and Robert G. Heath. "Septal Stimulation for the Initiation of Heterosexual Behavior in a Homosexual Male." *Journal of Behavior Therapy and Experimental Psychiatry* 3, no. 1 (March 1972): 23–30. https://doi.org/10.1016/0005-7916(72)90029-8.

Montagu, K. A. "Catechol Compounds in Rat Tissues and in Brains of Different Animals." *Nature* 180, no. 4579 (August 1957): 244–45. https://doi.org/10.1038/180244a0.

Montague, P. R., P. Dayan, and T. J. Sejnowski. "A Framework for Mesencephalic Dopamine Systems Based on Predictive Hebbian Learning." *Journal of Neuroscience* 16, no. 5 (March 1996): 1936–47. https://doi.org/10.1523/JNEUROSCI.16-05-01936.1996.

Moore, Omar Khayyam. "Divination—A New Perspective." *American Anthropologist* 59, no. 1 (February 1957): 69–74. https://www.jstor.org/stable/666530.

Moore, Talia Y., Kimberly L. Cooper, Andrew A. Biewener, and Ramanarayan Vasudevan. "Unpredictability of Escape Trajectory Explains Predator Evasion Ability and Microhabitat Preference of Desert Rodents." *Nature Communications* 8, no. 1 (2017): 440. https://doi.org/10.1038/s41467-017-00373-2.

Morgenstern, Oskar. "Perfect Foresight and Economic Equilibrium." In *Selected Economic Writings of Oskar Morgenstern*, edited by Andrew Schotter, 169–83. New York: New York University Press, 1976.

Moyer, Ann E. *The Philosophers' Game: Rithmomachia in Medieval and Renaissance Europe*. Ann Arbor: University of Michigan Press, 2001.

Nasar, Sylvia. *A Beautiful Mind*. New York: Simon & Schuster, 2011.

Nelson, D. Alan, Michael Wilson, and Lianne M. Kurina. "Post-traumatic Stress Disorder among U.S. Army Drone Operators." *Aerospace Medicine and Human Performance* 93, no. 7 (July 2022): 562–70. https://doi.org/10.3357/AMHP.6016.2022.

Nguyen, C. Thi. "Competition as Cooperation." *Journal of the Philosophy of Sport* 44, no. 1 (2017): 123–37. https://doi.org/10.1080/00948705.2016.1261643.

———. *Games: Agency As Art*. Oxford: Oxford University Press, 2020. https://doi.org/10.1093/oso/9780190052089.001.0001.

Nielsen, Michael. "Maps of Matter." *The Future of Matter* (blog), February 2021. https://futureofmatter.com/maps_of_matter.html.

Nijhuis, Michelle. "The Miracle of the Commons." *Aeon*, May 4, 2021. https://aeon.co/essays/the-tragedy-of-the-commons-is-a-false-and-dangerous-myth.

Noble, Denis. *The Music of Life: Biology Beyond Genes*. Oxford: Oxford University Press, 2008.

Norenzayan, Ara. *Big Gods: How Religion Transformed Cooperation and Conflict*. Princeton, NJ: Princeton University Press, 2013.

Norton, Michael I., and Samuel R. Sommers. "Whites See Racism as a Zero-Sum Game That They Are Now Losing." *Perspectives on Psychological Science* 6, no. 3 (May 2011): 215–18. https://doi.org/10.1177/1745691611406922.

O'Grady, Cathleen. "Fraudulent Data Raise Questions about Superstar Honesty Researcher." *Science*, August 24, 2021. https://www.science.org/content/article/fraudulent-data-set-raise-questions-about-superstar-honesty-researcher.

Olds, James. "Brain Stimulation and the Motivation of Behavior." In *Perspectives in Brain Research*, edited by M. A. Corner and D. F. Swaab, 401–26. Progress in Brain Research, vol. 45. Amsterdam: Elsevier, 1976. https://doi.org/10.1016/S0079-6123(08)61001-8.

———. "Physiological Mechanisms of Reward." In *Nebraska Symposium on Motivation, 1955*, edited by M. R. Jones, 73–139. Lincoln: University of Nebraska Press, 1955.

Olds, J., and P. Milner. "Positive Reinforcement Produced by Electrical Stimulation of Septal Area and Other Regions of Rat Brain." *Journal of Comparative and Physiological Psychology* 47, no. 6 (1954): 419–27. https://doi.org/10.1037/h0058775.

Oreskes, Naomi, and Erik M. Conway. *The Big Myth: How American Business Taught Us to Loathe Government and Love the Free Market*. New York: Bloomsbury, 2023.

Ostrom, Elinor. "A Behavioral Approach to the Rational Choice Theory of Collective Action: Presidential Address, American Political Science Association, 1997." *American Political Science Review* 92, no. 1 (1998): 1–22. https://doi.org/10.2307/2585925.

———. "Beyond Markets and States: Polycentric Governance of Complex Economic Systems." *American Economic Review* 100, no. 3 (2010): 641–72.

Ostrom, Vincent, and Elinor Ostrom. "Rethinking Institutional Analysis: Interviews with Vincent and Elinor Ostrom." By Paul Dragos Aligica. Mercatus Center, George Mason University, October 12, 2009. https://www.mercatus.org/research/research-papers/rethinking-institutional-analysis-interviews-vincent-and-elinor-ostrom.

Page, Scott E. *The Diversity Bonus: How Great Teams Pay Off in the Knowledge Economy*. Princeton, NJ: Princeton University Press, 2019.

Paley, William. *Natural Theology*. Oxford: Oxford University Press, 2008.

Parkinson, James. "An Essay on the Shaking Palsy." *Journal of Neuropsychiatry and Clinical Neurosciences* 14, no. 2 (May 2002): 223–36. https://doi.org/10.1176/jnp.14.2.223.

Partelow, Stefan, David J. Abson, Achim Schlüter, María Fernández-Giménez, Henrik von Wehrden, and Neil Collier. "Privatizing the Commons: New Approaches Need Broader Evaluative Criteria for Sustainability." *International Journal of the Commons* 13, no. 1 (2019): 747–76. https://doi.org/10.18352/ijc.938.

Pascal, Blaise. *Thoughts, Letters, and Minor Works*. Edited by Charles W. Eliot. Translated by W. F. Trotter, M. L. Booth, and O. W. Wight. New York: P. F. Collier & Son, 1910.

Patterson, William. "The Limits of 'Urban Dynamics.'" *Reason*, November 1971. https://reason.com/1971/11/01/the-limits-of-urban-dynamics.

Pearce, Jeremy. "Donald Michie, 83, Theorist of Artificial Intelligence, Dies." *New York Times*, July 23, 2007. https://www.nytimes.com/2007/07/23/science/23michie.html.

Peek, Philip M., ed. *African Divination Systems: Ways of Knowing*. Bloomington: Indiana University Press, 1991.

Penfield, Wilder. *No Man Alone: A Neurosurgeon's Life*. Boston and Toronto: Little, Brown and Company, 1977.

Perry, Tony. "Herbert York Dies at 87; Scientist and Arms-Control Leader." *Los Angeles Times*, September 29, 2014. https://www.latimes.com/nation/la-me-herbert-york21-2009may21-story.html.

Perry-Gal, Lee, Ian Stern, and Adi Erlich. "Gaming and Divination in the Hellenistic Levant: The Case Study of the Astragalus Assemblage from Maresha, Israel." *Levant* 54, no. 1 (2022): 65–79. https://doi.org/10.1080/00758914.2022.2048433.

Piaget, Jean. *The Moral Judgment Of The Child*. Translated by Marjorie Gabain. New York: Free Press, 1948.

Plato. *Plato: Complete Works*. Edited by John M. Cooper. Indianapolis: Hackett Publishing, 1997.

———. *Laws*. Translated by Benjamin Jowett. Urbana, IL: Project Gutenberg, 2008. https://www.gutenberg.org/files/1750/1750-h/1750-h.htm.

Plutarch. *De Stoicorum repugnantiis*. Edited by William W. Goodwin. Boston: Little, Brown and Company, 1874. http://www.perseus.tufts.edu/hopper/text?doc=Perseus:text:2008.01.0389:section=23.

Pope, Alexander. *The Works of Alexander Pope*. Philadelphia: Willis P. Hazard, 1856.

Popper, Karl, and John C. Eccles. *The Self and Its Brain: An Argument for Interactionism*. Berlin: Springer International, 1977.

Poundstone, William. *Prisoner's Dilemma*. New York: Anchor Books, 1993.

Price, George R. "The Nature of Selection." *Journal of Theoretical Biology* 175, no. 3 (August 1995): 389–96. https://doi.org/10.1006/jtbi.1995.0149.

Prochnik, George. *The Impossible Exile: Stefan Zweig at the End of the World*. New York: Other Press, 2014.

Queller, David C. "The Gene's Eye View, the Gouldian Knot, Fisherian Swords and the Causes of Selection." *Philosophical Transactions of the Royal Society B: Biological*

Sciences 375, no. 1797 (April 2020): 20190354. https://doi.org/10.1098/rstb.2019
.0354.

Quetelet, M. A. *A Treatise on Man and the Development of His Faculties.* Edinburgh:
William and Robert Chambers, 1842.

Radkau, Joachim. *Nature and Power: A Global History of the Environment.* Cambridge,
UK: Cambridge University Press, 2008.

Ramakrishna, Sri. *Selections from the Gospel of Sri Ramakrishna.* Annotated by Kendra
Crossen Burroughs. Translated by Swami Nikhilananda. Mumbai: Jaico Publishing
House, 2005.

Rapoport, Anatol. "The Use and Misuse of Game Theory." *Scientific American* 207, no.
6 (December 1962): 108–19. https://www.jstor.org/stable/24936389.

Rawls, John. *A Theory of Justice,* Rev. ed. Cambridge, MA: Harvard University Press,
1999.

Recorde, Robert. *The Whetstone of Witte: Whiche Is the Second Parte of Arithmetike;
Containyng Thextraction of Rootes: The Cossike Practise, with the Rule of Equation:
And the Woorkes of Surde Nombers.* London: J. Kyngstone, 1557.

Reeves, Richard. *President Kennedy: Profile of Power.* New York: Simon & Schuster,
2011.

Ridley, Mark, ed. *Last Words.* Vol. 3 of *Narrow Roads of Gene Land: The Collected Papers of W. D. Hamilton.* Oxford: Oxford University Press, 2006.

Rilke, Rainer Maria. *Letters to a Young Poet.* Translated by Joan M. Burnham. Novato,
CA: New World Library, 2000.

Robinson, Dan, and Georgios Konstantopoulos. "Ethereum Is a Dark Forest." Paradigm,
August 28, 2020. https://www.paradigm.xyz/2020/08/ethereum-is-a-dark-forest.

Robinson, Kim Stanley. *Pacific Edge: Three Californias.* New York: Tor, 1990.

Roeder, Oliver. *Seven Games: A Human History.* New York: W. W. Norton & Company, 2022.

Rollefson, Gary O. "A Neolithic Game Board from 'Ain Ghazal, Jordan." *Bulletin of the
American Schools of Oriental Research* 286 (May 1992): 1–5. https://doi.org/10.2307
/1357113.

Roper, Will. "AI Just Controlled a Military Plane for the First Time Ever." *Popular
Mechanics,* December 16, 2020. https://www.popularmechanics.com/military/
aviation/a34978872/artificial-intelligence-controls-u2-spy-plane-air-force
-exclusive.

Rosen, Robert. *Life Itself: A Comprehensive Inquiry into the Nature, Origin, and Fabrication of Life.* New York: Columbia University Press, 1991.

Ross, Don. "Game Theory." In *The Stanford Encyclopedia of Philosophy,* edited by Edward N. Zalta. Stanford, CA: Stanford University, 2021. https://plato.stanford.edu
/archives/fall2021/entries/game-theory/.

Rota, Gian-Carlo. *Indiscrete Thoughts.* Edited by Fabrizio Palombi. Boston: Birkhäuser,
1997.

Roth, Alvin E., Tayfun Sönmez, and M. Utku Ünver. "A Kidney Exchange Clearinghouse in New England." *American Economic Review* 95, no. 2 (May 2005): 376–80.
https://doi.org/10.1257/000282805774669989.

———. "Pairwise Kidney Exchange." Working Paper 10698, National Bureau of Economic Research, Cambridge, MA, August 2004. https://doi.org/10.3386/w10698.

Roy, Pradipto. "Global Pharma and Local Science: The Untold Tale of Reserpine." *Indian Journal of Psychiatry* 60, no. S2 (February 2018): S277–83. https://doi.org/10.4103/psychiatry.IndianJPsychiatry_444_17.

Różycka-Tran, Joanna, Jarosław P. Piotrowski, Magdalena Żemojtel-Piotrowska, Paweł Jurek, Evgeny N. Osin, Byron G. Adams, Rahkman Ardi, et al. "Belief in a Zero-Sum Game and Subjective Well-Being across 35 Countries." *Current Psychology* 40, no. 7 (2021): 3575–84. https://doi.org/10.1007/s12144-019-00291-0.

Rubinstein, Ariel. *Economic Fables*. Cambridge, UK: Open Book Publishers, 2012.

———. "The Best Books on Game Theory." Interview by Sophie Roell. Five Books (website), December 6, 2016. https://fivebooks.com/best-books/ariel-rubinstein-on-game-theory.

Rumi. *The Essential Rumi*. Translated by Coleman Barks with John Moyne. Edison, NJ: Castle Books, 1997.

Rushdie, Salman. *Midnight's Children*. New York: Random House, 2006.

Sacks, Oliver. *Awakenings*. New York: Knopf Doubleday, 2013.

Sambursky, S. "On the Possible and the Probable in Ancient Greece." *Osiris* 12 (1956): 35–48.

Samek, Anya, Jason M. Cowell, Alexander W. Cappelen, Yawei Cheng, Carlos Contreras-Ibáñez, Natalia Gomez-Sicard, Maria L. Gonzalez-Gadea, et al. "The Development of Social Comparisons and Sharing Behavior across 12 Countries." *Journal of Experimental Child Psychology* 192 (April 2020): 104778. https://doi.org/10.1016/j.jecp.2019.104778.

Samuel, Arthur L. "Programming Computers to Play Games." In *Advances in Computers*, edited by Franz L. Alt, 165–92. Vol. 1. New York and London: Academic Press, 1960. https://doi.org/10.1016/S0065-2458(08)60608-7.

Samuelson, Paul A. "A Note on Measurement of Utility." *Review of Economic Studies* 4, no. 2 (February 1937): 155–61. https://doi.org/10.2307/2967612.

Saramago, José. *All the Names*. Translated by Margaret Jull Costa. New York: Harcourt, 1999.

Schaeffer, Jonathan. *One Jump Ahead: Computer Perfection at Checkers*. Rev. ed. New York: Springer, 2009. https://doi.org/10.1007/978-0-387-76576-1.

Schell, Jesse. *The Art of Game Design: A Book of Lenses*. Amsterdam and Boston: Elsevier / Morgan Kaufmann, 2008.

Schelling, Thomas C. *Arms and Influence*. New Haven, CT: Yale University Press, 1966.

———. *Arms and Influence*. New Haven, CT: Yale University Press, 2020.

———. *The Strategy of Conflict*. Cambridge, MA: Harvard University Press, 1980.

Schiller, Friedrich. *On the Aesthetic Education of Man*. Translated by Reginald Snell. Mineola, NY: Dover Publications, 2012.

Schlesinger, Keith. "SimBill, SimHillary: The High Price of Medi-Carelessness." *Computer Gaming World*, June 1994.

Schotter, Andrew, ed. *Selected Economic Writings of Oskar Morgenstern*. New York: New York University Press, 1976.

Schüll, Natasha Dow. *Addiction by Design: Machine Gambling in Las Vegas*. Princeton, NJ: Princeton University Press, 2012. https://doi.org/10.1515/9781400834655.

Schultz, W., P. Apicella, and T. Ljungberg. "Responses of Monkey Dopamine Neurons to Reward and Conditioned Stimuli during Successive Steps of Learning a Delayed Response Task." *Journal of Neuroscience* 13, no. 3 (March 1993): 900–913. https://doi.org/10.1523/JNEUROSCI.13-03-00900.1993.

Schultz, Wolfram. "Dopamine Reward Prediction Error Coding." *Dialogues in Clinical Neuroscience* 18, no. 1 (March 2016): 23–32.

Schultz, Wolfram, Peter Dayan, and P. Read Montague. "A Neural Substrate of Prediction and Reward." *Science* 275, no. 5306 (March 1997): 1593–99. https://doi.org/10.1126/science.275.5306.1593.

Schultz, Wolfram, Regina M. Carelli, and R. Mark Wightman. "Phasic Dopamine Signals: From Subjective Reward Value to Formal Economic Utility." *Current Opinion in Behavioral Sciences* 5 (October 2015): 147–54. https://doi.org/10.1016/j.cobeha.2015.09.006.

Schweitzer, Albert. *Indian Thought and Its Development.* Boston: Beacon Press, 1957.

———. *Out of My Life and Thought: An Autobiography.* Translated by Antje Bultmann Lemke. Baltimore: Johns Hopkins University Press, 1998.

Scott, James C. *Seeing Like a State: How Certain Schemes to Improve the Human Condition Have Failed.* New Haven, CT: Yale University Press, 1999.

Seeger, Raymond J. "Von Neumann, Jewish Catholic." *Perspectives on Science and Christian Faith* 40 (December 1988): 234–36. https://www.asa3.org/ASA/PSCF/1988/PSCF12-88Seeger.html.

Sejnowski, Terrence. "Dopamine Made You Do It." In *Think Tank*, edited by David J. Linden, 257–62. New Haven, CT: Yale University Press, 2018. https://doi.org/10.12987/9780300235470-034.

Sen, Amartya K. "Rational Fools: A Critique of the Behavioral Foundations of Economic Theory." *Philosophy & Public Affairs* 6, no. 4 (Summer 1977): 317–44. https://www.jstor.org/stable/2264946.

Shannon, Claude E. "XXII. Programming a Computer for Playing Chess." *London, Edinburgh, and Dublin Philosophical Magazine and Journal of Science* 41, no. 314 (March 1950): 256–75. https://doi.org/10.1080/14786445008521796.

Shenk, David. *The Immortal Game: A History of Chess.* New York: Anchor Books, 2007.

Sherwin, Martin J. "One Step from Nuclear War." *Prologue*, Fall 2012. https://www.archives.gov/publications/prologue/2012/fall/cuban-missiles.html.

Speck, Frank G. *Naskapi: The Savage Hunters of the Labrador Peninsula.* Norman, OK: University of Oklahoma Press. 1977.

Shubik, Martin. "Game Theory and Operations Research: Some Musings 50 Years Later." Working Paper Series ES, Working Paper No. 14, Yale School of Management, New Haven, CT, May 2001. https://doi.org/10.2139/ssrn.271029.

———. "Game Theory at Princeton, 1949–1955: A Personal Reminiscence." In *Toward a History of Game Theory*, edited by E. Roy Weintraub, 151–64. Durham, NC: Duke University Press, 1992.

Shultz, Susanne, and R. I. M. Dunbar. "The Evolution of the Social Brain: Anthropoid Primates Contrast with Other Vertebrates." *Proceedings of the Royal Society B: Biological Sciences* 274, no. 1624 (October 2007): 2429–36. https://doi.org/10.1098/rspb.2007.0693.

Sidak, J. Gregory. *Foreign Investment in American Telecommunications.* Chicago: University of Chicago Press, 1997.

Simon, Herbert A. *The Sciences of the Artificial*. Reissued 3rd ed. Cambridge, MA: MIT Press, 2019.

Skinner, B. F. *Beyond Freedom and Dignity*. Indianapolis: Hackett Publishing, 2002.

——. *Contingencies of Reinforcement: A Theoretical Analysis*. Cambridge, MA: B. F. Skinner Foundation, 2014.

——. "'I Have Been Misunderstood': An Interview with B. F. Skinner." *Center Magazine*, March/April 1972.

Skyrms, Brian. *Evolution of the Social Contract*. 2nd ed. Cambridge, UK: Cambridge University Press, 2014.

Smith, Adam. *The Theory of Moral Sentiments*. 2nd ed. London: A. Millar, 1761.

Sommer, Elisabeth W. "Gambling with God: The Use of the Lot by the Moravian Brethren in the Eighteenth Century." *Journal of the History of Ideas* 59, no. 2 (April 1998): 267–86. https://doi.org/10.2307/3653976.

Speers, Lucinda J., and David K. Bilkey. "Maladaptive Explore/Exploit Trade-Offs in Schizophrenia." *Trends in Neurosciences* 46, no. 5 (May 2023): 341–54. https://doi.org/10.1016/j.tins.2023.02.001.

Spinka, Marek, Ruth C. Newberry, and Marc Bekoff. "Mammalian Play: Training for the Unexpected." *Quarterly Review of Biology* 76, no. 2 (July 2001): 141–68.

Spinoza, Baruch. *The Essential Spinoza: Ethics and Related Writings*. Edited by Michael L. Morgan. Translated by Samuel Shirley. Indianapolis: Hackett Publishing, 2006.

Sridam, Nattaphan, and Kammant Phanthumchinda. "Encephalitis Lethargica–like Illness: Case Report and Literature Review." *Journal of the Medical Association of Thailand* 89, no. 9 (September 2006): 1521–27.

Srinivasan, Ashwin, ed. *Donald Michie: Machine Intelligence, Biology and More*. Oxford: Oxford University Press, 2009.

St. George, Donna. "Crown Him. He Can Beat a Computer. The Unsung Champion of Checkers." *Philadelphia Inquirer*, April 11, 1993.

Staňková, Kateřina, Joel S. Brown, William S. Dalton, and Robert A. Gatenby. "Optimizing Cancer Treatment Using Game Theory: A Review." *JAMA Oncology* 5, no. 1 (2019): 96–103. https://doi.org/10.1001/jamaoncol.2018.3395.

Starr, Paul. "Seductions of Sim: Policy as a Simulation Game." *American Prospect*, April 1, 1994. https://prospect.org/environment/seductions-sim-policy-simulation-game.

Stein, Stephen W., and Charles G. Thiel. "The History of Therapeutic Aerosols: A Chronological Review." *Journal of Aerosol Medicine and Pulmonary Drug Delivery* 30, no. 1 (2017): 20–41. https://doi.org/10.1089/jamp.2016.1297.

Steinmetz, Andrew. *The Gaming Table: Its Votaries and Victims*. Vol. 1. London: Tinsley Brothers, 1870.

Sterling, Bruce. *Zeitgeist*. New York: Bantam Books, 2000.

Stevens, Matthew W. R., Diana Dorstyn, Paul H. Delfabbro, and Daniel L. King. "Global Prevalence of Gaming Disorder: A Systematic Review and Meta-Analysis." *Australian & New Zealand Journal of Psychiatry* 55, no. 6 (June 2021): 553–68. https://doi.org/10.1177/0004867420962851.

The State of Texas, et al. v. Google LLC, No. CIVIL NO. 4:20-CV-957-SDJ, 2021. https://www.texasattorneygeneral.gov/sites/default/files/images/admin/2020/Press/20201216_1%20Complaint%20(Redacted).pdf.

Strathern, Marilyn. "'Improving Ratings': Audit in the British University System." *European Review* 5, no. 3 (July 1997): 305–21. https://doi.org/10.1002/(SICI)1234 -981X(199707)5:3<305::AID-EURO184>3.0.CO;2-4.

Strauss, Lewis L. *Men and Decisions*. New York: Doubleday, 1962.

Sugden, Robert. "The Community of Advantage." *Economic Affairs* 39, no. 3 (October 2019): 417–23. https://doi.org/10.1111/ecaf.12374.

Suits, Bernard. *The Grasshopper: Games, Life and Utopia*. Toronto: University of Toronto Press, 1978.

Sutton, Richard, and Andrew Barto. *Reinforcement Learning: An Introduction*. Adaptive Computation and Machine Learning Series. 2nd ed. Cambridge, MA: MIT Press, 2018.

Sutton, Richard. "Episode 11—Richard Sutton." Interview by Craig Smith. *Eye on A.I.* (podcast), March 20, 2019. MP3 audio, 38:08. https://www.eye-on.ai/podcast-011.

———. "Fourteen Declarative Principles of Experience-Oriented Intelligence." *Incomplete Ideas* (blog), 2008. http://incompleteideas.net/RLAIcourse2009/princi ples2.pdf.

———. "The Bitter Lesson." *Incomplete Ideas* (blog), March 13, 2019. http://www.in completeideas.net/IncIdeas/BitterLesson.html.

———. "Verification, the Key to AI." *Incomplete Ideas* (blog), November 15, 2001. http://incompleteideas.net/IncIdeas/KeytoAI.html.

Tägil, Sven. "Alfred Nobel's Thoughts about War and Peace." Nobel Prize (website), November 20, 1998. www.nobelprize.org/alfred-nobel/alfred-nobels-thoughts -about-war-and-peace.

Taleb, Nassim Nicholas. *Fooled by Randomness: The Hidden Role of Chance in Life and in the Markets*. New York: Random House, 2016.

Tesauro, Gerald. "Temporal Difference Learning and TD-Gammon." *Communications of the ACM* 38, no. 3 (March 1995): 58–68. https://doi.org/10.1145/203330 .203343.

Thompson, Ken. "Oral History of Ken Thompson." Interview by John Mashey. Computer History Museum, February 8, 2005. Transcript. http://archive.computerhis tory.org/resources/text/Oral_History/Thompson_Ken/thompson.oral_history _transcript.2005.102657921.pdf.

Thorndike, E. L. "Animal Intelligence: An Experimental Study of the Associative Processes in Animals." *Psychological Review: Monograph Supplements* 2, no. 4 (1898): i—109. https://doi.org/10.1037/h0092987.

———. *The Elements of Psychology*. New York: A. G. Seiler, 1905.

Tik, Martin, Ronald Sladky, Caroline Di Bernardi Luft, David Willinger, André Hoffmann, Michael J. Banissy, Joydeep Bhattacharya, and Christian Windischberger. "Ultra-high-field FMRI Insights on Insight: Neural Correlates of the Aha! Moment." *Human Brain Mapping* 39, no. 8 (August 2018): 3241–52. https://doi.org /10.1002/hbm.24073.

Toole, Betty Alexandra, ed. *Ada, the Enchantress of Numbers: A Selection from the Letters of Lord Byron's Daughter and Her Description of the First Computer*. Gloucestershire, UK: Strawberry Press, 1992.

Turing, A. M. "Computing Machinery and Intelligence." *Mind* 59, no. 236 (1950): 433–60.

———. "Digital Computers Applied to Games." In *Faster than Thought: A Symposium on Digital Computing Machines*, edited by B. V. Bowden, 286–310. London: Pitman & Sons, 1953.

Turkle, Sherry. "Seeing Through Computers." *American Prospect*, December 19, 2001. https://prospect.org/api/content/67843f8b-cfa7-530c-8b89-97ed0136e3ed.

Twain, Mark. *Autobiography of Mark Twain*. Vol. 1, edited by Harriet Elinor Smith. Berkeley: University of California Press, 2010.

Ulam, Stanislaw. *Adventures of a Mathematician*. Berkeley: University of California Press, 1991.

———. "John von Neumann 1903–1957." *Bulletin of the American Mathematical Society* 64, no. 3 (1958): 1–49. https://doi.org/10.1090/S0002-9904-1958-10189-5.

Ullmann-Margalit, Edna. *The Emergence of Norms*. Oxford: Oxford University Press, 2015.

van den Herik, Jaap. "Computer Chess Today and Tomorrow: An Interview with Donald Michie." *Computer Chess Digest Annual*, 1984.

Vardi, Moshe Y. "Artificial Intelligence: Past and Future." *Communications of the ACM* 55, no. 1 (January 2012): 5. https://doi.org/10.1145/2063176.2063177.

Vego, Milan. "German War Gaming." *Naval War College Review* 65, no. 4 (2012): 106–48.

von Hilgers, Philipp. *War Games: A History of War on Paper*. Cambridge, MA: MIT Press, 2012.

von Mises, Ludwig. *Economic Calculation in the Socialist Commonwealth*. Translated by S. Alder. Ludwig von Mises Institute, 1990. https://mises.org/library/economic-calculation-socialist-commonwealth/html.

von Neumann, John, and Oskar Morgenstern. *Theory of Games and Economic Behavior*. 60th anniversary commemorative ed. Princeton, NJ: Princeton University Press, 2007.

von Neumann, John. "Can We Survive Technology?" *Fortune*, June 1955.

———. "Defense in Atomic War." *Ordnance* 40, no. 216 (1956): 1090–92. https://www.jstor.org/stable/45360902.

———. Letter to Oswald Veblen, May 21, 1943. Oswald Veblen Papers, Box 15. Library of Congress, Washington, DC.

———. "Nomination of John von Neumann to Be a Member of the United States Atomic Energy Commission," March 8, 1955. John von Neumann Papers, Library of Congress, Washington, DC.

———. "On the Theory of Games of Strategy." In *Contributions to the Theory of Games*, edited by A. W. Tucker and R. D. Luce, 13–42. Vol. 4. Princeton, NJ: Princeton University Press, 1959. https://doi.org/10.1515/9781400882168-003.

———. *The Neumann Compendium*. Edited by F. Bródy and T. Vamos. World Scientific Series in 20th-Century Mathematics, vol. 1. Singapore: World Scientific, 1995.

———. "Theory of Games I (General Foundations)," 1940. Oskar Morgenstern Papers, Special Collections Department, Perkins Library, Duke University, Durham, NC.

Vonneuman, Nicholas A. "John von Neumann as Seen By His Brother." Unpublished manuscript, 1987. https://www.math.ru.nl/~mueger/vonneumann.pdf.

Warneken, Felix, and Michael Tomasello. "The Emergence of Contingent Reciprocity in Young Children." *Journal of Experimental Child Psychology* 116, no. 2 (October 2013): 338–50. https://doi.org/10.1016/j.jecp.2013.06.002.

Warner, Silas. *RobotWar.* Baltimore: Muse Software, 1981. https://corewar.co.uk/ro botwar/robotwar.txt.

Watson, Patty. *Archaeological Ethnography in Western Iran.* Tucson, AZ: University of Arizona Press, 1979.

Weber, Bruce. "A Mean Chess-Playing Computer Tears at the Meaning of Thought." *New York Times,* February 19, 1996. https://archive.nytimes.com/www.nytimes .com/partners/microsites/chess/archive8.html.

Weinstein, Michael M. "Bringing Logic to Bear on Liberal Dogma." *New York Times,* December 1, 2002. https://www.nytimes.com/2002/12/01/weekinreview/the-na tion-bringing-logic-to-bear-on-liberal-dogma.html.

Weintraub, E. Roy. *Toward a History of Game Theory.* Durham, NC: Duke University Press, 1992.

Wells, H. G. *Little Wars.* London: Frank Palmer, 1913.

Weyl, E. Glen. "How Market Design Economists Helped Engineer a Mass Privatization of Public Resources." *ProMarket* (blog), May 28, 2020. https://www.promarket.org /2020/05/28/how-market-design-economists-engineered-economists-helped -design-a-mass-privatization-of-public-resources.

Weyl, Hermann. "The Current Epistemological Situation in Mathematics." In *From Brouwer to Hilbert: The Debate on the Foundations of Mathematics in the 1920s,* edited by Paolo Mancosu, 123–42. Oxford: Oxford University Press, 1998.

Wigner, Eugene P. "Two Kinds of Reality." *Monist* 48, no. 2 (1964): 248–64.

———. *The Collected Works of Eugene Paul Wigner.* New York: Springer Science & Business Media, 2001.

Wintjes, Jorit. "Europe's Earliest Kriegsspiel? Book Seven of Reinhard Graf zu Solms' Kriegsregierung and the 'Prehistory' of Professional War Gaming." *British Journal for Military History* 2, no. 1 (November 2015): 15–33.

———. "When a Spiel Is Not a Game." *Vulcan* 5, no. 1 (2017): 5–28. https://doi.org /10.1163/22134603-00501002.

Yager, David D. "Predator Detection and Evasion by Flying Insects." *Current Opinion in Neurobiology* 22, no. 2 (April 2012): 201–7. https://doi.org/10.1016/j.conb.2011 .12.011.

Yamamoto, Kei, and Philippe Vernier. "The Evolution of Dopamine Systems in Chordates." *Frontiers in Neuroanatomy* 5 (March 2011): 21. https://www.frontiersin .org/articles/10.3389/fnana.2011.00021.

Yeragani, Vikram K., Manuel Tancer, Pratap Chokka, and Glen B. Baker. "Arvid Carlsson, and the Story of Dopamine." *Indian Journal of Psychiatry* 52, no. 1 (2010): 87–88. https://doi.org/10.4103/0019-5545.58907.

Yoo, Cheong-mo. "Go Master Lee Says He Quits, Unable to Win over AI Go Players." *Yonhap News Agency,* November 27, 2019. https://en.yna.co.kr/view/AEN201911 27004800315.

York, Herbert. *Race to Oblivion: A Participant's View of the Arms Race.* New York: Simon & Schuster, 1970.

Zahavy, Tom, Vivek Veeriah, Shaobo Hou, Kevin Waugh, Matthew Lai, Edouard Leurent, Nenad Tomasev, Lisa Schut, Demis Hassabis, and Satinder Singh. "Diversifying AI: Towards Creative Chess with AlphaZero." Preprint, submitted August 17, 2023. https://doi.org/10.48550/arXiv.2308.09175.

Zermelo, Ernst. "Über eine Anwendung der Mengenlehre auf die Theorie des Schachspiels." In *Proceedings of the Fifth International Congress of Mathematicians*, edited by E. W. Hobson and A. E. H. Love, 501–4. Vol. 2. Cambridge, UK: Cambridge University Press, 1913.

Zhang, Jingsong, Jessica J. Cunningham, Joel S. Brown, and Robert A. Gatenby. "Integrating Evolutionary Dynamics into Treatment of Metastatic Castrate-Resistant Prostate Cancer." *Nature Communications* 8, no. 1 (November 2017): 1816. https://doi.org/10.1038/s41467-017-01968-5.

Zweig, Friderike. *Married to Stefan Zweig*. Lexington, MA: Plunkett Lake Press, 2019.

Zweig, Stefan. *The Collected Novellas of Stefan Zweig*. Translated by Anthea Bell. London: Pushkin Press, 2021.

Index